Macmillan Encyclopedia of the Environment

Macmillan Encyclopedia of the
ENVIRONMENT

VOLUME 5

General Editor
Stephen R. Kellert

Associate Editors
Matthew Black

Richard Haley

Macmillan Library Reference USA
New York

Developed, Designed, and Produced by Book Builders Incorporated

Macmillan Library Reference
1633 Broadway, New York, NY 10019-6785

Library of Congress Catalog Card Number: 96-29045

Printed in the United States of America

Library of Congress Cataloging-in-Publication Data

Macmillan encyclopedia of the environment.
     p.   cm.
     "General editor, Stephen R. Kellert"—P. iii.
     Includes bibliographical references and index.
     Summary: Provides basic information about such topics as minerals, energy resources, pollution, soils and erosion, wildlife and extinction, agriculture, the ocean, wilderness, hazardous wastes, population, environmental laws, ecology, and evolution.
     ISBN 0-02-897381-X (set)
     1. Environmental sciences—Dictionaries, Juvenile.
    [1. Environmental protection—Dictionaries. 2. Ecology—Dictionaries.]
    I. Kellert, Stephen R.   96-29045
    GE10.M33 1997   CIP
    333.7—dc20   AC

Photo credits are gratefully acknowledged in a special listing in Volume 6, page 102.

P-Q continued

Pesticide

▶ Substance used to kill unwanted organisms. The term *pesticides* includes FUNGICIDES, RODENTICIDES, HERBICIDES, and INSECTICIDES. Pesticides may be used to protect humans or other organisms from disease-carrying or disease-causing organisms. Chemical pesticides are widely used on farms to protect crops. They are also used on lawns to kill destructive INSECTS and unwanted PLANTS.

The use of pesticide is not new. Ancient Roman farmers dusted their crops with diatomaceous earth (pulverized shells of tiny sea creatures called *diatoms*) to kill insect pests. Today, people are dependent on chemical pesticides. Such use has increased crop production and has helped in controlling diseases such as malaria which is transmitted by mosquitoes.

DRAWBACKS TO PESTICIDE USE

There are several drawbacks to the use of pesticides. Among them are AIR POLLUTION and WATER POLLUTION. When pesticides are applied by planes called crop dusters, approximately 5% of the pesticide reaches its target. The other 95% may enter the air or nearby water. For example, the toxic pesticide atrazine is sometimes found in high concentrations in the air in Maryland.

Another drawback of pesticides is that rain washes pesticides from cropland into SURFACE WATER and groundwater. In many Midwestern agricultural areas, people are advised to drink bottled water because their wells contain high levels of pesticides. FISH may also be contaminated by pesticides in surface water. Such fish are often not safe to eat.

Some pesticides remain in the ENVIRONMENT for long periods of time. These pesticides can poison people and WILDLIFE years after use. Pesticides may also spread widely—even globally. For example, Antarctic penguins have been found to contain the pesticide dichlorodiphenyl trichloroethane (DDT) and other pesticides in their body tissues. The United States stopped the use of DDT in 1972. Many other nations of the world discontinued use of DDT soon after.

Some chemical pesticides may be passed through FOOD CHAINS and accumulate in the body tissues of organisms in a process called BIOACCUMLATION. Such pesticides may

Some Common Pesticides		
Types	**Uses**	**Characteristics**
Acidimine	Herbicide	Inhibits germination
Arsenic compounds	Herbicide, insecticide, rodenticide	
Carbamates	Insecticide	Persist for days
Chlorinated hydro-carbons	Insecticides such as DDT and hydrocarbons, chlordane	Persist for years
Dinitro compounds	Fungicide, insecticide, herbicide	
Mercury	Fungicide	Extremely toxic
Organophosphates	Insecticide	Persist for weeks
Phenoxy	Herbicides such as 2,4-D and 2,4,5,-T	
Phosphorus	Rodenticide	Extremely toxic
Pyrethroids	Insecticide	Persist for days
Thiocarbamates	Herbicide	Inhibit germination
Triazines	Herbicide	Inhibit photosynthesis

eventually cause health problems in organisms or their offspring. Chemical pesticides also kill beneficial insects, such as plant pollinators, as well as destructive pests. For example, pesticides used on Indonesian rice fields in the 1980s killed the destructive brown planthopper. They also killed the planthopper's natural enemies. In a few years, planthopper populations developed a resistance to pesticides, and rice production plummeted.

One of the gravest problems arising from chemical pesticide use is pest resistance. Some individual pests have GENES that enable them to survive pesticides. These individuals mate and pass along their pesticide-resistant genes. Through the process of NATURAL SELECTION, subsequent generations are born with a natural resistance to pesticides.

In the 1950s, about 30 SPECIES of insect pests were immune to chemical pesticides. By 1990, 504 insect and mite crop pests were immune to all known pesticides. In the United States, pesticide use increased tenfold between 1945 and 1989. Though we annually use 500,000 tons (450,000 metric tons) of pesticides (costing $4.1 billion), crop losses from insect damage have increased from 7% to 13%.

Many pesticides cause diseases and illnesses in people. Of the 600 pesticides used in the United States, only 41 have been tested for toxic health effects. However, CANCER rates among farmers in the agricultural Midwest are generally higher than the national average.

ALTERNATIVES TO CHEMICAL PESTICIDES

Some farmers have reduced their dependence on chemical pesticides by implementing a system of INTEGRATED PEST MANAGEMENT (IPM). IPM uses BIOLOGICAL CONTROLS to combat pests. For example, pesticides are made from PLANTS having natural insecticides instead of from chemicals.

Another IPM method is the introduction of natural PREDATORS to kill pests. However, the introduction of EXOTIC SPECIES as predators has sometimes caused more harm than good. This was shown in the case of the introduction of mongooses to control rat populations in West Indian islands and Hawaii. The rats were able to avoid being eaten by the mongooses by climbing trees. To keep from starving, the mongooses fed on the poultry and the wild native birds in the islands.

Users of IPM avoid monoculture planting, in which a single crop species is planted, year after year, on the same ground. MONOCULTURE invites pest infestations. Instead, farmers use CROP ROTATION in which different plants are grown on different fields each year with some fields periodically left unplanted, or fallow, to allow them to regenerate. [*See also* ADAPTATION; AGRICULTURAL POLLUTION; AGROECOLOGY; CARSON, RACHEL LOUISE; DEFOLIANT; and GENETIC ENGINEERING.]

◆ Used pesticide barrels that are improperly discarded pose a threat to the groundwater.

Petrochemical

▶A chemical product, such as synthetic rubber, PLASTIC, DETERGENT, vinyl, and nylon, that is made using NATURAL GAS or PETROLEUM. The raw materials used for making petrochemicals consist of HYDROCARBONS, compounds that contain only hydrogen and CARBON. Hydrocarbons, such as ethylene, acetylene, and propylene, are known as first-stage petrochemicals. They are produced synthetically at oil refineries. When combined with other substances, first-stage petrochemicals can be used to produce many products that are useful to humans.

PETROCHEMICAL PRODUCTS

Petrochemical products are used in nearly every industry from agriculture to medicine. Ethylene is perhaps the most widely used hydrocarbon in the petrochemical industry. It is used to make anti-freeze. It is also important in the production of synthetic rubber, latex paints, and polyethylene—a transparent plastic. Acetylene is another important hydrocarbon for industry. Obtained from METHANE, acetylene is either burned directly to produce the heat energy used in welding or is used to make plastics, adhesives, and vinyl.

ENVIRONMENTAL PROBLEMS

Petrochemical plants discharge many polluting chemicals into the ATMOSPHERE, including SULFUR DIOXIDE. They also produce solid, nonbiodegradable wastes that are placed in LANDFILLS. In landfills, these wastes release toxic chemicals into SOIL and groundwater.

Gases released during the manufacture of petrochemicals can be toxic to humans. In 1984, toxic fumes accidentally released from a pesticide plant in Bhopal, India, killed 2,500 people. PESTICIDES are petrochemical products used in agriculture.

Accidents such as the one in Bhopal are rare. The chemical industry, in general, has one of the best safety records of all industries. However, the accident does illustrate the potential dangers of petrochemical products.

Many environmental laws currently regulate the POLLUTION produced by petrochemical plants. However, there will always be a trade-off between the environmental dangers produced by petrochemicals and the need of society to use the products they are used to make. [*See also* BHOPAL INCIDENT and BIODEGRADABLE.]

◆ Petrochemical plants process natural gas and petroleum, producing chemicals used in nearly every industry.

Petroleum

▶A FUEL commonly called *crude oil*. The word *petroleum* comes from the Latin words meaning *rock* and *oil*. The people who first found it observed the substance seeping from cracks in surface rocks.

Geologists believe petroleum formed from the remains of organisms, primarily PLANKTON and simple PLANTS, which lived millions of years ago. This theory is based on the presence of materials in oil that could have come only from organisms. The link between petroleum and organisms of long ago has led to the description of petroleum as a FOSSIL FUEL.

◆ The first oil well in the United States was erected at Titusville, Pennsylvania, in 1859.

FORMATION OF PETROLEUM

Remains of organisms were buried in fine-grained SEDIMENT at the bottom of OCEANS that covered much of Earth's surface millions of years ago. As the sediment deepened, the remains were subjected to higher temperatures and pressures. Eventually, the remains were compressed into sedimentary rock.

Through chemical processes, the rock formed a waxy substance called *kerogen*. When temperatures rose above 400° F (204° C), the kerogen separated into a liquid (oil) and a gas (NATURAL GAS).

Over long periods of time, the oil and natural gas moved upward through openings in and between rocks and collected in porous reservoir rocks—chiefly sandstone and

◆ Oil spilled by the *Exxon Valdez* washes ashore, covering a dead duck, algae, and rocks.

limestone. The oil and gas moved again within the reservoir rocks to collect in pools. The pools of oil were generally capped by waterproof layers of rock and soil.

HISTORY OF PETROLEUM USE

Thousands of years ago, some people began making use of petroleum that seeped through Earth's surface. The Sumerians used ASPHALT, a form of petroleum, as an adhesive

when making jewelry and statues. Pitch, a solid form of petroleum, was also used by the Egyptians to coat mummies.

Hundreds of years before Europeans arrived in the Americas, Native Americans used crude oil as fuel and as medicine. In the 1600s, missionaries traveling through what is now Pennsylvania discovered Native Americans scooping oil from surface pools. Pioneers heading West bought oil from Kit Carson to use as axle grease.

◆ The coal made from these ancient trees may be turned into gasoline by a process called *coal liquefaction*.

THE PETROLEUM INDUSTRY

Most historians trace the beginning of the petroleum industry to 1869. In this year, Edwin Drake drilled the first oil well near Titusville, Pennsylvania. Following the drilling of this first well, the petroleum industry grew quickly. Geologists began discovering larger oil deposits in other states. By the late 1800s, major drilling sites for oil began operation in Kentucky, Ohio, Indiana, and Illinois.

In the United States, annual oil production jumped from 2,000 barrels in 1859 to 64 million barrels in 1900. In 1901, the first North American gusher was produced at Spindletop Field in eastern Texas. Soon after this event, California and Oklahoma also became leading oil-producing states.

Today, the major oil-producing regions of the world are located in the Middle East, the United States, Russia, Africa, and Venezuela. The oil produced from these areas supplies about half the world's energy. In addition, the oil provides the raw materials for a variety of PETROCHEMI-CALS, which are chemicals from petroleum.

RECOVERING OIL

Geological and geophysical studies usually indicate places where petroleum might have accumulated. However, there is only a 2% chance

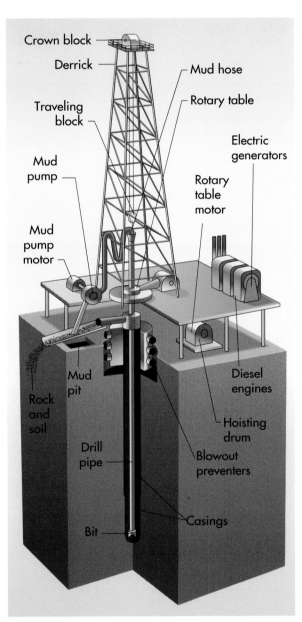

◆ Drilling rigs include an above-ground derrick and machinery for raising and lowering equipment. Below ground, the bit drills through the soil and rock to the oil below.

that the petroleum is present in useful amounts. Often, many **dry holes** are drilled before oil begins to flow from a productive well.

In the United States, the rotary drilling technique is used to recover oil from beneath Earth's surface. A drill bores through the ground. Pressurized mud is pumped down through pipes to carry bits of rock and soil to the surface and to prevent the sudden release of pressure in the reservoir of oil below. When the well is completed, the oil rises to the surface under the natural pressure in the reservoir.

If not enough pressure is in the reservoir to bring oil to the surface, artificial means may be used. One way of completing this task is to inject water into the reservoir. The water displaces the oil, forcing it to flow up into the well. Pumps on Earth's surface keep the pressure going.

REFINING OIL

After crude oil reaches the surface, the natural gas that collects in the well is separated from the liquid oil. The gas is sent to processing plants or directly to consumers. Salt, water, and sediment are removed from the oil. The oil then travels by pipelines to storage tanks, tankers, or refineries. Pipelines can move millions of barrels of oil each day. Refineries differ in size and capacities.

At a refinery, oil is separated by **fractional distillation** to make products such as gasoline, kerosene, diesel oil, fuel oil, lubricating oil, and asphalt. Small refineries process about 150 barrels of oil per day. Some large refineries process more than 600,000 barrels of oil each day.

PETROLEUM AND THE ENVIRONMENT

Fires, explosions, and OIL SPILLS are hazards associated with recovering and transporting oil. Pipelines may erode, allowing oil to leak onto the ground. Such oil threatens the plant and animal life living in the area. Oil tankers have spilled millions of barrels of crude oil into OCEANS, bays, lakes, and rivers. Much of the spilled oil floats on top of the water where it is carried by waves and OCEAN CURRENTS to shorelines. Such oil spills harm organisms living in the water and on the shorelines. For example, sea birds and marine animals are often covered in gooey oil mixture. The oil may make it impossible for these organisms to move, breathe, or carry out other life processes needed for their survival. Other animals that prey on oil-covered organisms may die from ingesting the oil.

As petroleum products are used by consumers, environmental hazards continue. Gasoline and oil products that are not disposed of properly can contaminate groundwater. In addition, HYDROCARBONS and wastes formed from burning petroleum and its products that are released into the atmosphere create environmental and health problems. [*See also* ALASKA PIPELINE; ALTERNATIVE ENERGY SOURCES; CATALYTIC CONVERTERS; EXXON VALDEZ; GASOHOL; GREENHOUSE EFFECT; OIL DRILLING; OIL POLLUTION; OIL SHALE; ORGANIZATION OF PETROLEUM EXPORTING COUNTRIES (OPEC); PARTICULATES; and SCRUBBER.]

◆ At refineries like this one, oil is processed to make products such as gasoline, lubricating oil, and asphalt.

Pet Trade

❿The buying, selling, or swapping of rare or exotic animals throughout the world. Almost all human cultures keep pets. People love pets for many reasons—for their beauty, devotion, and companionship. Dogs, cats, hamsters, and goldfish are popular pets in the United States. In most cases, these types of pets are not sold or traded between countries. They are sold to pet stores by local breeders or donated by owners. Other more exotic types of pets, such as mon-

◆ Exotic animals, such as iguanas, are part of the pet trade.

keys, iguanas, parrots, turtles, frogs, tropical FISH, corals, and sea anemones, are imported from other nations and sold legally by pet stores throughout the world. These types of animals are often admired for their novelty. Frequently, they are used as status symbols.

DRAWBACKS TO THE PET TRADE

The buying and selling of exotic pets is a huge business. However, there are drawbacks to the pet trade. Among these drawbacks are the illegal HUNTING practices used to obtain some SPECIES and the damage done to ECOSYSTEMS when animals are removed from an area.

Illegal Hunting

Like any product that is bought and sold, a rare animal is more valuable than a common animal. Pet owners are sometimes willing to pay thousands of dollars for an exotic species. Such high prices often lead to the illegal buying and selling of animals, even if the animals are protected by international laws. For example, certain species of monkeys and parrots are commonly bought and sold illegally. Conservationists are concerned about the POACHING and trading of rare species because many of them are ENDANGERED SPECIES which are threatened with EXTINCTION.

Rare species that can bring high prices often are captured illegally, smuggled out of the country as hidden cargo, then sold to individuals as pets. This problem goes beyond the illegal, black market trade. Legal pet dealers may become unknowingly involved in the problem when they purchase pets that have been illegally captured.

Threats to Ecosystems

A major drawback to the exotic pet trade is the damage to ecosystems that sometimes occurs when animals are collected. Such damage occurs on both land and water ENVIRONMENTS, but it is most devastating when coral reef animals are collected. CORAL REEFS are among the most diverse ecosystems in the world. A variety of organisms—including fish, eels, crabs, shrimp, turtles, sea stars, sea anemones, and jellyfish—make their homes in coral reefs. When corals are collected, portions of the reef are destroyed in the process. This temporarily disrupts the HABITAT of countless species.

MONITORING THE PET TRADE

In the United States, the ENDANGERED SPECIES ACT of 1973 is the main body

of law that controls the types of animals that can be captured and traded. The Endangered Species Act prohibits the capturing and killing of all endangered and threatened species. On an international level, the poaching and illegal trading of WILDLIFE is controlled by the CONVENTION ON THE INTERNATIONAL TRADE OF ENDANGERED SPECIES OF WILD FAUNA AND FLORA (CITES). CITES consists of more than 100 member countries. Members of CITES meet every other year to determine which species require international protection. [*See also* ANIMAL RIGHTS; HABITAT LOSS; INTERNATIONAL UNION FOR THE CONSERVATION OF NATURE (IUCN); WILDLIFE CONSERVATION; and WILDLIFE MANAGEMENT.]

◆ Most rare species that become pets are captured and sold legally. However, many exotic pets that are rare and protected by law are hunted and sold illegally on the black market.

◆ This Philippine eagle, the mynahs, and parrots have been captured for sale in the pet trade.

◆ A pH meter measures hydrogen ion concentration.

pH

▶ A measure of the hydrogen ions, or charged particles, in a liquid. The more hydrogen ions there are in a liquid, the more acidic it will be.

The measurement of pH uses a mathematical system called *logarithms* to express the number of hydrogen ions in simple terms. The pH scale extends from 0 to 14. On this scale, any number below 7 is considered acidic. The lower the number, the stronger the acid. Any number above 7 is considered alkaline, or basic; the higher the number the stronger the alkali. Seven is neutral; it is neither acidic nor alkaline.

The pH scale is important to life on Earth because many chemical processes on which life depends occur only at a specific pH level. For example, if the pH of a person's arterial blood does not stay within the range of 7.0 to 7.8, the person will die. [*See also* ACID RAIN.]

Phosphate

▶ Any of a variety of chemical compounds containing the chemical phosphorus. Phosphorus occurs in nature as phosphate. Phosphate is contained in rocks, animal wastes, and in the tissues of living organisms. It is an important chemical for the growth and proper functioning of most living things. However, many environmentalists are concerned about human activities that add excess phosphorus in the form of phosphates to the ENVIRONMENT.

Like other nutrients, phosphorus cycles naturally through ECOSYSTEMS. PLANTS obtain phosphorus from the SOIL. In turn, when animals eat plants or other animals, they obtain the phosphorus that originated in plants. When organisms die and decompose, phosphorus is returned to the soil. Once in the soil the phosphorus can once again be absorbed by plant roots.

Phosphorus is a nonliving or ABIOTIC FACTOR in the environment. The amount of phosphorus in the environment places limits on the sizes of the populations of plants and ALGAE the environment can support. The amount of phosphorus is particularly important in aquatic environments. As the amount of phosphorus in an aquatic HABITAT increases as a result of EROSION and RUNOFF from roads and feedlots, plant and algae populations thrive. In healthy, undisturbed ecosystems, phosphorus levels generally remain in balance.

Scientists are concerned about human contributions to the phosphorus cycle. Excess phosphorus in the environment can lead to dramatic population explosions of

◆ Massive fish kills may result from too much phosphate in a lake or pond.

algae and plants. In aquatic ecosystems, such population explosions can eventually lead to EUTROPHICATION. Population explosions involving algae and plants result in a rapid loss of DISSOLVED OXYGEN in the water. Too little OXYGEN in water disrupts ecosystems by suffocating plants and animals, particularly FISH and larval AMPHIBIANS.

REDUCING PHOSPHATE POLLUTION

In the past, household DETERGENTS containing phosphates were prime contributors to phosphate pollution. Phosphates have been banned from these products. However, they still enter the environment in WASTEWATER and when runoff from farms washes chemical fertilizers into lakes and rivers.

Several methods are used to remove excess phosphorus from wastewater. Primary and secondary SEWAGE treatments are generally ineffective at removing phosphorus. Thus, special procedures at the third or tertiary stage of sewage treatment must be used to remove the extra phosphorus. One method involves the use of BACTERIA that absorb phosphorus from wastewater. This technique results in the production of phosphorus-rich SLUDGE, which can be separated from wastewater and removed.

A more popular method of phosphorus removal uses chemicals that react with phosphorus. The chemicals cause phosphorus to form a solid that can be separated from wastewater. These treatments are very effective and remove about 95% of the phosphorus originally present in wastewater. [*See also*

FOOD CHAIN; FOOD WEB; INDUSTRIAL WASTE TREATMENT; MINERALS; WASTEWATER, PRIMARY, SECONDARY, AND TERTIARY TREATMENT OF; and SEWAGE TREATMENT PLANT.]

Photodegradable Plastics

▶ PLASTICS that break down when exposed to certain kinds of light. Since the development of plastics began in the 1920s, the number and types of products that have been created using these materials have changed the lives of people world-

wide. However, one of the main characteristics of plastics—their ability to remain intact long after they are still useful to people—also poses a hazard to the ENVIRONMENT.

Unlike products made from natural materials, such as paper and wood, most plastics do not degrade, or break down, over time. As a result, many plastic products that have been buried in LANDFILLS remain intact years after their disposal.

Two methods for dealing with the accumulation of plastics in landfills are being used. The first involves recycling, the reprocessing of plastics for use in other products. This method for reducing the amounts of plastics that are disposed of in landfills is being used for products such as discarded beverage containers. The other method

◆ Over time, photodegradable plastics break apart when exposed to sunlight. However, such plastics break only into smaller pieces; they do not change chemically.

being used to combat the plastics problem is the creation of degradable plastics. Such plastics are made to break apart when disposed of in the environment.

There are two types of degradable plastics—photodegradable plastics and BIODEGRADABLE plastics. Both types of plastics are used to make items such as trash or GARBAGE bags or shopping bags used by stores and supermarkets. Photodegradable plastics are made to break apart when exposed to sunlight for several weeks or longer. Under such conditions, these plastics become brittle and break apart into smaller pieces. Biodegradable plastics are made using special chemicals that cause the large molecules that make up the plastics to break down into smaller molecules over time. In addition, the plastics contain some natural materials, such as cornstarch, that are designed to attract microorganisms. When placed in landfills, BACTERIA and other DECOMPOSERS feed upon the natural materials contained in the plastic. As the bacteria feed, they make tiny holes in the plastic material. Over time, the combined action—the breaking down of plastic molecules and the holes made by the bacteria—cause the plastic to break into smaller pieces of plastic.

The use of the term *degradable* in the names of plastics that break into smaller pieces is a bit misleading. Most often this term is applied to materials that break down chemically in the environment. However, degradable plastics do not break down chemically; they break down physically. Thus, the plastic does not change in composition; it changes only in size. As a result, large pieces of plastic placed into landfills do not change into substances that are recycled naturally in the environment. Instead, the plastics change only into smaller pieces of plastic. [*See also* BIOGEO-CHEMICAL CYCLE; CHEMICAL CYCLES; and DECOMPOSITION.]

Photosynthesis

❯❯The process by which PLANTS make carbohydrates using OXYGEN and water. Photosynthesis means "making with light." A short way of showing photosynthesis is:

$$CO_2 + H_2O + \text{light energy} \rightarrow$$
(carbon dioxide) + (water)
$$CH_2O + O_2$$
(carbohydrates) + (oxygen)

Photosynthesis is carried out in plants, blue-green BACTERIA, and some protists.

About 350 years ago, Belgian physician J. B. Van Helmont grew a willow tree in a pot of SOIL for five years. He measured the weight gained by the tree and discovered that it was much greater than the weight lost by the soil in the pot. This was a mystery: if most of the material in the plant was not made from soil, where was it coming from?

In 1771, English chemist Joseph Priestley discovered that plants could "improve" air. When a candle was burned in a closed jar, the air became "bad" and unable to support a mouse in the jar. When a plant was in the jar, the air sup-

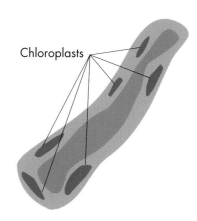

Chloroplasts

Leaf cell
with chloroplasts

◆ Photosynthesis occurs in chloroplasts, which are most abundant in plant leaves.

ported the breathing of the mouse. Dutch physician Jan Ingenhousz later discovered that the plant supplies air only if it is in sunlight. Here was another mystery: what did a plant add to the air, and why did it only happen in the light? Since then, the work of many more scientists has provided the answers to these questions.

HOW PHOTOSYNTHESIS WORKS

Plants use the energy of sunlight to chemically combine water and CARBON DIOXIDE, which is present in the

◆ Every leaf is like a solar panel. Inside the chloroplasts of a plant, energy from sunlight is used to make food.

air. This reaction produces oxygen and **carbohydrates.** There are many kinds of carbohydrates. For example sugars, starches, and the plant material called *cellulose* are all carbohydrates. Cellulose makes up the cell walls of plants. Plants are therefore made mostly of materials taken from the air and water. This chemical reaction also explains why a plant could save the life of a mouse in one of Priestley's closed jars. The mouse needed oxygen to breathe and the plant provided the oxygen. The plant also took in the carbon dioxide the mouse produced when it breathed out.

Photosynthesis does not occur exactly the same way in every kind of plant. Many plants live in water, so they get carbon dioxide from the water instead of the air. Other organisms, such as blue-green bacteria or "cyanobacteria," carry on photosynthesis without having chloroplasts—the structures that often hold the chlorophyll that cap-

tures the sun's energy. However, all photosynthetic organisms use the sun's energy to set up a flow of **electrons,** or a tiny electrical current, to power the process of making carbohydrates.

WHY PHOTOSYNTHESIS IS IMPORTANT

The ATMOSPHERE is about 20% oxygen as a result of photosynthesis. The presence of oxygen in the atmosphere is one of the most important changes that living things have made on this planet. It would be impossible for most modern plants and animals to live without oxygen in the air. They depend on one form of oxygen (O_2) for breathing. They depend on another form of oxygen (O_3, or OZONE) for protection against the harmful ULTRAVIOLET RADIATION given off by the sun.

Photosynthesis also removes carbon dioxide from the air. Today, extra carbon dioxide is added to the air by human activities, such as the burning of FOSSIL FUELS. The carbon dioxide in the air helps regulate temperatures on Earth by trapping heat near Earth's surface. Scientists theorize that GLOBAL WARMING may occur as carbon dioxide levels increase. Growing more plants could remove some of this extra carbon dioxide. However, this is only one of many actions needed to reduce a buildup of carbon dioxide in the atmosphere. [*See also* BIOGEOCHEMICAL CYCLE; CARBON CYCLE; CHEMICAL CYCLES; CLIMATE CHANGE; DEFORESTATION; FOOD CHAIN; GREENHOUSE EFFECT; GREENHOUSE GAS; GREEN REVOLUTION; PRODUCER; and RESPIRATION.]

THE LANGUAGE OF THE ENVIRONMENT

carbohydrates compounds made of carbon, hydrogen, and oxygen atoms.

electrons Particles in an atom that orbit around the nucleus and carry a negative charge. The movement of electrons produces electricity.

Photovoltaic Cell

I A device that converts light energy, usually from the sun, into ELECTRICITY. Photovoltaic cells are also known as *solar cells*. These cells are most often used to convert SOLAR ENERGY into electricity. The photovoltaic cell is "environmentally friendly" because it does not use up Earth's NONRENEWABLE RESOURCES to generate power.

Photovoltaic cells are used in the exposure meters of cameras. In a camera, a photovoltaic cell measures the intensity of the light source. These cells are used to power machines such as handheld calculators, watches, and toys. Photovoltaic cells have also been used for AUTOMOBILES, airplanes, and artificial satellites sent into space. The use of photovoltaic cells in transportation vehicles is intended to reduce AIR POLLUTION caused by the FOSSIL FUELS that normally power these machines. Photovoltaic cells are also used to detect electromagnetic RADIATION. [*See also* ALTERNATIVE ENERGY SOURCES; CONSERVATION; and ENERGY EFFICIENCY.]

◆ Photovoltaic cells convert energy from the sun or other light source into electric energy that can be used to power machines such as this boat.

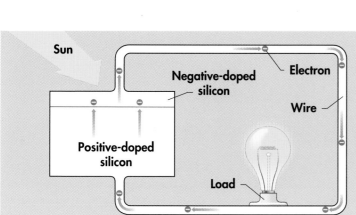

◆ The internal mechanism of a photovoltaic cell is shown here.

Phytoplankton

I Tiny plantlike organisms that float near the surface of the ocean and freshwater ECOSYSTEMS. The name *phytoplankton* means "plants that wander." These **microscopic** organisms float from place to place in aquatic HABITATS, such as ponds, lakes, WETLANDS, and the oceans.

Phytoplankton are small ALGAE made up of only a single cell or a few cells. Like other algae, phytoplankton carry out PHOTOSYNTHESIS. Although phytoplankton contain chlorophyll, many of these organisms are not green. Instead, they may be brown, gold, or red due to the presence of pigments in their cells. Although they are small, phytoplankton are present in huge numbers and are very important in the world's aquatic ecosystems.

THE IMPORTANCE OF PHYTOPLANKTON

Like plants, phytoplankton use energy from sunlight to make food. In any ecosystem, organisms that carry out photosynthesis provide food for other organisms that cannot photosynthesize. During the day, when most photosynthesis occurs, phytoplankton take in CARBON DIOXIDE from the ATMOSPHERE and release OXYGEN back to the atmosphere. Thus, phytoplankton help regulate the amounts of carbon dioxide and oxygen in the air and water around them.

The largest ecosystem supported by phytoplankton is the open OCEAN. Phytoplankton may be the only PRODUCERS present in areas of thousands of square miles. Small animal-like organisms called ZOOPLANKTON feed on the phytoplankton. These organisms are then eaten by larger animals, such as FISH, BIRDS, and marine MAMMALS. All these zooplankton and animals depend on a food supply that begins with marine phytoplankton. Thus, phytoplankton form the base of aquatic FOOD CHAINS. The importance of phytoplankton is not limited only to

marine animals. A large portion of Earth's oxygen, which we breathe, is also produced by marine phytoplankton.

THE IMPORTANCE OF NUTRIENTS

Phytoplankton need more than just sunlight, water, and carbon dioxide to grow. Like other living things, they also need nutrients, such as nitrogen, phosphorus, and iron. Some of the best places for phytoplankton to grow are where rivers spill nutrients into the ocean and where OCEAN CURRENTS bring up nutrients from the sea floor. In many other parts of the ocean, nutrients are in short supply, so phytoplankton do not grow abundantly there.

Some people believe it might be possible to add nutrients to

some parts of the ocean to raise more phytoplankton. Having more phytoplankton might help people in two ways. More phytoplankton can feed more fish, which people could harvest and eat. More phytoplankton might also help to remove more carbon dioxide from the atmosphere and help slow GLOBAL WARMING.

Scientists have added nutrients to lakes and to small portions of the ocean, resulting in increased growth of phytoplankton. However, more experiments need to be done

◆ Phytoplankton are not all the same. Just like the plants in a forest or field, these tiny ocean plants come in many sizes, shapes, and colors and belong to different species.

to decide whether fertilizing phytoplankton in large areas of ocean is really a good idea.

PHYTOPLANKTON IN DANGER

Like all other forms of aquatic life, phytoplankton can be affected by POLLUTION. Sometimes the pollutant is a nutrient that permits too much growth of phytoplankton. For example, phosphate fertilizers that wash from farmland into lakes can cause an ALGAL BLOOM. In other cases, a pollutant may be a **toxin**, such as a PESTICIDE or PETROLEUM product.

Another threat to phytoplankton is ozone depletion in Earth's atmosphere. The thinning of the OZONE LAYER allows more of the sun's ULTRAVIOLET RADIATION to reach the surface waters, where phytoplankton live. Overexposure to ultraviolet radiation has been shown to be harmful to many organisms. It is not yet known whether too much exposure will harm phytoplankton and decrease the food supply in aquatic food chains, nor even how much exposure is too much. [*See also* ALGAE; CARBON CYCLE; DETERGENTS; DISSOLVED OXYGEN; FOOD WEB; MARINE POLLUTION; and PHOSPHATE.]

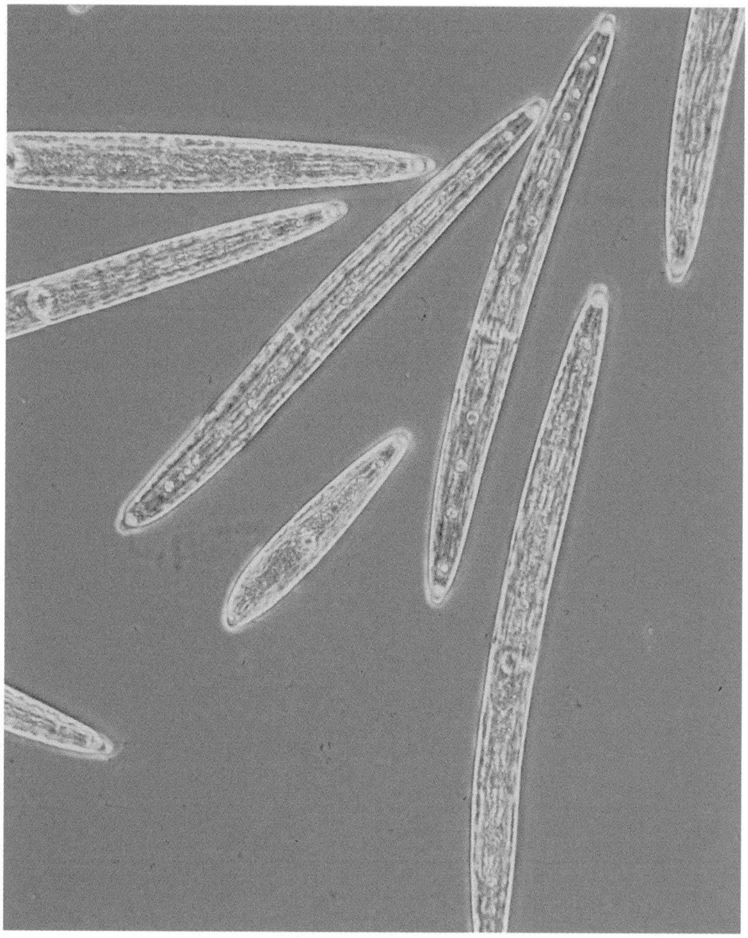

◆ Green algae provide a large percentage of the oxygen in the atmosphere.

Pinchot, Gifford (1865–1946)

First chief of the U.S. FOREST SERVICE from 1898 to 1910. In 1905, when the U.S. Forest Service became a separate agency in the DEPARTMENT OF AGRICULTURE, Gifford Pinchot was appointed head by President Theodore ROOSEVELT.

Gifford Pinchot was born in Connecticut on August 11, 1865, at the end of the Civil War. After graduating from Yale University, he went abroad to study FORESTRY in France and Switzerland. Upon his return to the United States, Pinchot set about developing methods of efficient and profitable timber production. He encouraged selective cutting. He also encouraged thinning FORESTS of dead and diseased wood, scrub growth, and mature trees. He established the principle of "constant annual yield" and instructed forest crews in methods

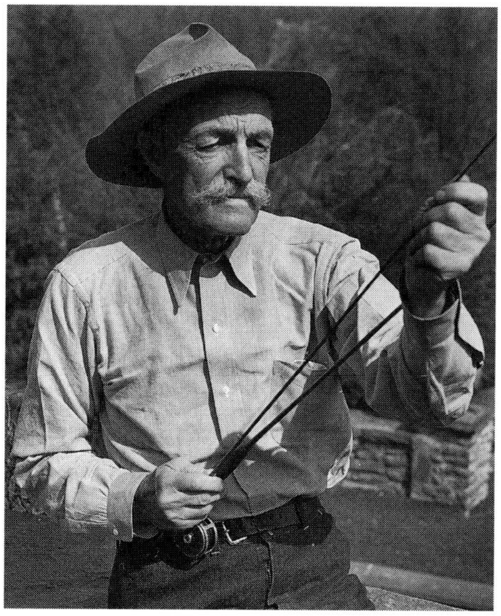

of cutting trees without destroying the forests.

Pinchot argued that the reason for saving forests was not for their beauty or to provide shelter for animals but because of their use in providing homes and jobs for people. He, along with Theodore Roosevelt, fought against waste and corruption, and for scientific, modern, professional management of resources for the greatest good of the people. Pinchot believed that carefully managed forests could continue to produce valuable timber indefinitely. These principles continue to guide the United States Forest Service today.

In 1904, Pinchot became embroiled in a bitter controversy with his long-time friend, John MUIR, when Pinchot supported the flooding of the Hetch-Hetchy Valley in YOSEMITE NATIONAL PARK. In the early 1900s, San Francisco wanted to dam the Tuolumne River to produce HYDROELECTRIC POWER and provide water for the city. Muir opposed this idea because it would flood the Hetch-Hetchy Valley, destroying its beauty and grandeur. Pinchot supported the plan because he felt that unused land is wasteland. After a prolonged, bitter fight, Muir lost the battle and the DAM was built.

In 1910, Pinchot lost his position as the nation's chief forester. He was fired by Roosevelt's successor, President Taft. Pinchot clashed with Secretary of the Interior Richard Ballinger over WILDERNESS areas in Montana and Wyoming. These areas had been set aside as protected federal lands during Theodore Roosevelt's presidency. Then, in mid-1909, Ballinger reopened the lands for sale to private interests. When Pinchot wrote a letter to Congress calling Ballinger's actions dangerous, President Taft fired him for disloyalty. Pinchot also served as a professor at the Yale University School of Forestry, which he helped found, and served two terms as the governor of Pennsylvania.

Pioneer Species

The first groups of organisms to populate any newly available area and begin the process of ecological SUCCESSION. Pioneer species are typically hardy, rapidly reproducing organisms, such as weeds, mosses, and LICHENS. Such SPECIES are associated with primary succession, the process in which ECOSYSTEMS develop in previously uninhabited areas.

Some examples of areas likely to undergo primary succession include newly created islands and the areas exposed when GLACIERS retreat. Primary succession is also common at construction sites, in sidewalk cracks, and on the walls of old brick buildings.

The colonization of an area by pioneer species represents the first stage in ecological succession. Pioneer species begin ecosystem development by making conditions suitable for other species. This process often includes the development of SOIL in areas in which soil was not present. An example of the changes that occur during primary succession may be observable on a city sidewalk. Pioneer species that can colonize this type of ENVIRONMENT include rapidly growing and reproducing organisms, such as weeds, mosses, and FUNGI. Mosses and weeds may take root in microscopic cracks in the sidewalk. As they grow, they break apart the concrete in the sidewalk into small particles. As these species die, BAC-

◆ Hardy, rapidly reproducing liverworts and mosses make up part of the community of pioneer species.

♦ Mosses, lichens, and fungi are the first to populate a newly available area.

TERIA and other DECOMPOSERS break them down. In this process, nutrients are added to the growing pile of soil. Over time, enough fertile soil may be formed to support the growth of more complex plants, such as grasses and shrubs.

Plankton

❙Small organisms that float near the surfaces of OCEANS or lakes. If we tow a fine net through the ocean, it will trap marine plankton, many of which are so small they can be seen only through a microscope. Many plankton are green or yellow ALGAE, plantlike protists that make food by PHOTOSYNTHESIS. The algae include diatoms, microscopic organisms with shells like glass, and dinoflagellates, plantlike protists that have whiplike appendages that propel them through the water. Because algae need light for photosynthesis, they live only near the surface where light can penetrate the water. Photosynthetic plankton are called PHYTOPLANKTON. *Phytoplankton* produce much of the OXYGEN present in the air.

Phytoplankton are common wherever the ocean is rich in the mineral nutrients needed for photosynthesis and reproduction. These nutrients are found near continents, where rivers wash MINERALS from the land into the oceans. In some parts of the open ocean, upwelling currents bring nutrients from the ocean bottom to the surface. Here too, concentrations of plankton are found.

Plankton make up the main food for larger ocean animals, such as FISH. Thus, fishing is best near land and near upwellings, ocean water that churns up nutrients from the bottom. Satellites now photograph the ocean and identify areas where lots of plankton and schools of fish are to be found.

Diatoms and dinoflagellates first appeared on Earth many millions of years ago. Some of the oldest known fossils are of diatoms. Diatom fossils are found everywhere on Earth.

Some organisms that live by eating phytoplankton are also plankton. Such organisms are called ZOOPLANKTON. Many zooplankton form shells of calcium carbonate. When these organisms die, their shells may sometimes accumulate in great numbers. The White Cliffs of Dover in England are made of limestone formed from zooplankton shells.

Zooplankton include animallike protists called *protozoa*. Like phytoplankton, most zooplankton are microscopic. Some, however, are large enough to be seen with the unaided eye.

Although plankton are not strong enough to swim against OCEAN CURRENTS, many can move up and down in the water. Currents near the surface often travel in dif-

♦ Plankton include these small algae.

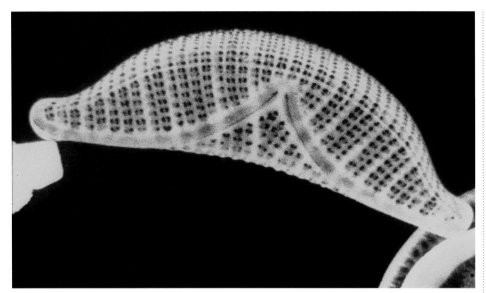

◆ The type of plankton called diatoms comes in many shapes and sizes.

ferent directions from those deeper down. This ability of zooplankton and phytoplankton to move deeper and shallower in water permits them to move from one current to another and travel vast distances.

Plankton live in freshwater lakes and streams as well as in the ocean. However, freshwater plankton are not nearly as diverse as marine plankton. One reason for this lack of diversity is that few freshwater animals produce larvae. Many of the larvae in lakes and streams are insect larvae. Most fresh water flows into the sea. If a freshwater clam or crayfish produced larva, the larvae would be washed out to sea, where it would be killed by the SALINITY. Instead, most freshwater animals produce eggs that hatch into organisms resembling miniature adults. Often those organisms are equipped with hooks or suckers that enable them to cling to rocks and mud and avoid being washed away.

Plant Pathology

▶The study of PLANT diseases caused by organisms, such as worms, BACTERIA, and FUNGI. VIRUSES are also common plant PATHOGENS.

Healthy plants have a natural ability to resist diseases. Thus, most plant diseases occur when plants do not get enough sunlight, water, or nutrients. Disease also occurs quickly when a plant has been damaged by INSECTS or weather-related factors such as wind or snow.

Farmers use many techniques to control plant diseases. Traditional methods include the use of PESTICIDES to limit damage caused by insects, weeds, rodents, and fungi. Fertilizers are used to keep SOIL supplied with essential plant nutrients. However, because many such farming practices also pose risks to the ENVIRONMENT, many farmers are turning to SUSTAINABLE AGRICULTURE and ORGANIC FARMING techniques to keep plants free from disease. Such methods limit environmental damage. For example, to improve the quality of soil, farmers use CROP ROTATION and NO-TILL AGRICULTURE. Such farming methods add nutrients to the soil without the use of fertilizers. Scientists are also using biotechnology to develop new types of plants that are better able to resist damage by pathogens and insects. [*See also* BIOLOGICAL CONTROL; FUNGICIDE; GENETIC ENGINEERING; GREEN REVOLUTION; INSECTICIDE; INTEGRATED PEST MANAGEMENT (IPM); PEST CONTROL; and RODENTICIDE.]

Plants

▶The kingdom made of unicellular and multicellular (many-celled) organisms that make their own food through the process of PHOTOSYNTHESIS. Plants include some ALGAE, as well as FERNS, pines, and FLOWERING PLANTS. Plants provide us with food; wood for building and newspapers; COAL, oil, and NATURAL GAS used for transportation and making ELECTRICITY; and the OXYGEN we breathe.

Plants range in size from the microscopic single-celled green algae to the gigantic redwood trees. Almost all plants are photosynthetic—they make their own food from sunlight and CARBON DIOXIDE, using chloroplasts, the tiny cellular structures containing chlorophyll. A few plant species have evolved to become PARASITES or DECOMPOSERS, and no longer use photosynthesis. Plant cells, unlike most animal cells, have cell walls. Plants have a variety of shapes, ranging from seaweeds to mosses, trees, and flowering plants.

THE LIFE CYCLE OF PLANTS

There are two kinds of plant adults—the sporophyte, which has a complete set of chromosomes, and the haploid adult, which has half the number of chromosomes.

The sporophyte divides to produce spores, which develop into haploid adults. The female haploid adult produces eggs and the male produces sperm. The sperm is produced after the process of POLLINATION, after which it fertilizes the egg. The union of a sperm and an egg of two haploid adult plants forms a diploid plant, whose reproductive cells divide to form a haploid spore. In this way the cycle continues in which the diploid and haploid generations alternate.

TYPES OF PLANTS

There are several types of plants. They are the algae, bryophytes, and the vascular plants—gymnosperms and angiosperms.

Algae

There are three kinds of algae—the brown, red, and green algae. Aside from their color, these types of algae differ in the type of chemical they use to store food reserves. Most algae are multicellular. Many are microscopic, but a few, such as

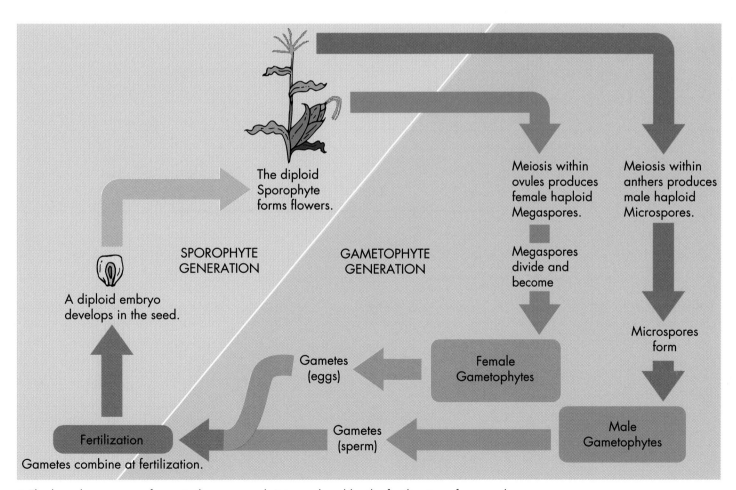

The diploid Sporophyte forms flowers.

SPOROPHYTE GENERATION

GAMETOPHYTE GENERATION

Meiosis within ovules produces female haploid Megaspores.

Meiosis within anthers produces male haploid Microspores.

Megaspores divide and become

Microspores form

A diploid embryo develops in the seed.

Gametes (eggs)

Female Gametophytes

Gametes (sperm)

Male Gametophytes

Fertilization

Gametes combine at fertilization.

◆ The kernel on an ear of corn is the corn seed. It is produced by the fertilization of an egg by a sperm.

kelps, are quite large. None have true roots, stems, and leaves, although some, like the red and brown algae, look like they do. The internal structure of algae is vastly different from the internal structure of more complex plants.

The brown and red algae are marine plants. They live in the OCEANS, either attaching themselves to something solid or floating free in the water. Green algae live on land as well as in water. All three types of algae help to form the bases of aquatic FOOD CHAINS.

Bryophytes

Approximately 420 million years ago, an ancestral stock of green algae gave rise to the first land plants called *bryophytes*. Bryophytes are nonvascular plants—

◆ Plants have many roles in ecosystems. They provide food, oxygen, and shelter for animals, and their roots hold the soil in place.

◆ Sunflowers are angiosperms whose seeds are covered with a protective envelope.

they have no complex internal system for circulating water and providing support. Some examples of bryophytes are mosses, liverworts, and hornworts. These are all small plants that live in moist environments. Bryophytes have evolved a waxy covering, an ADAPTATION that allows them to survive on land without drying out, as algae do.

The sperm of a bryophyte reaches the egg through puddles of rainwater, which is one reason that bryophytes do not grow tall. The female plants have to remain close to the ground for the sperm to reach the eggs.

Vascular Land Plants

Vascular land plants are plants that have an internal skeleton made of tissues that support the plant body in an upright position and function as a transport system. There are two kind of vascular tissue—xylem and phloem. Xylem is a major supporting tissue that transports water and MINERALS from the roots to all parts of the plant. Xylem is the woody part of the plant. Phloem transports the sap that contains nutrients produced through photosynthesis to all parts of the plant.

Vascular plants reproduce in a number of ways. The early vascular plants which still exist today, such as ferns, horsetails, and club moss, reproduce like bryophytes—a sperm reaches an egg through puddles. The later vascular plants are the gymnosperms, such as pines, spruces, firs, and redwoods, and the angiosperms, also called FLOWERING PLANTS. Gymnosperms and angiosperms have seeds, structures that contain and protect the

embryo plant, and can withstand adverse environmental conditions, such as low temperatures and lack of rainfall.

Gymnosperms evolved about 350 million years ago. They have needlelike or scalelike leaves and are large plants with cones or conelike structures. The tallest tree measures 370 feet high (112 meters) and the largest measures 102 feet (31 meters) in diameter at its base. The largest gymnosperms are the redwoods of California. The oldest known individual tree is a bristlecone pine in Inyo National Forest in California. This tree is about 4,600 years old.

Because of the height of some gymnosperms, reproduction often occurs in treetops. Pollen from male cones float through the air. Some land on female cones and grow into them, producing the male haploid adult that makes sperms, which fertilizes the eggs in the young seeds in the female cone. The zygote, or the new individual, develops into an embryo inside the seed.

Angiosperms, or flowering plants, are the best known and most abundant of all plant groups. They are widely distributed in all HABITATS and range in size from the tiny, floating duckweeds to the giant oaks. Angiosperms are the most recent group of plants to evolve, appearing 160 million years ago. [*See also* ALGAL BLOOM; AUTOTROPH; PHYTOPLANKTON; SOFTWOODS; and SILVICULTURE.]

◆ Kelp, a brown alga, forms vast forests under the ocean.

Plastic

▶ A synthetic material made up of long chains of molecules called *polymers*. Each chain is formed by hundreds or even millions of links of small repeating molecules called *monomers*. The monomers making up plastics consist of CARBON, hydrogen, nitrogen, and OXYGEN atoms.

Some polymers are a series of just one kind of monomer. Others have several kinds of monomers in alternating patterns or in a random sequence. Some chains are firm; others are elastic, depending on the kinds of monomers and the arrangement of their molecules. The different arrangements of molecules give plastic its most important characteristic—the ability to be formed into almost any shape.

KINDS OF PLASTIC

Plastics are classified as either thermosettings (thermosets) or thermoplastics, according to how they behave when heated. Thermosets can be heated and shaped only once because heating causes crosslinking. Crosslinking is a chemical reaction that binds the polymer chains together. Once hardened, thermosets cannot be made liquid again. This characteristic makes thermosets useful in making heat-resistant items such as the pot handles and trays that are used for sterilizing medical instruments.

Thermoplastics can be melted and reshaped over and over again. Heating does not bind their polymer chains in a permanent arrangement. Molecules in the chains move freely each time the thermoplastics are heated. This characteristic makes thermoplastics easy to work with. Thermoplastics also require less time to set than thermosets—about 10 seconds, compared to 5 minutes for thermosets. The moving molecules of thermoplastics can sometimes make them lose their shape under continual pressure. Thermosets are commonly used for making chairs and seats for buses.

THE HISTORY OF PLASTIC

Plastic was accidentally invented in 1907 by Leo Baeleland, a Belgian-American chemist. Baeleland was trying to create an extremely strong and durable shellac, a varnish used to give wood a shiny finish. Instead, he made a clear-material that was hard, lightweight, and easily shaped. He called his creation Bakelite and used it for making unbreakable dishes. Before the invention of plastic, the scientist John Wesley Hyatt worked with a similar material called *celluloid*.

In the years that followed Baeleland's discovery, scientists in Germany, England, and the United States perfected many other plastics—each with its own distinct characteristics. Four important thermoplastics—acrylics, nylon, polystyrene, and polyvinyl chloride (PVC or vinyl)—became available in the 1930s. In the 1940s, polyester, polyethylene, silicone, and epoxy (a gluelike susbstance) resins were developed. In 1953, General Motors Corporation introduced the Chevrolet Corvette—the first mass-produced AUTOMOBILE with a polyester body.

◆ Today, millions of different products are made from plastic, one of the most important inventions of the twentieth century.

In the 1960s, rapid growth of the PETROCHEMICAL industry, the chief producer of raw materials for plastics, also increased plastic production. Today, people use millions of plastic products, including car bumpers, baby bottles, toys, milk containers, bread wrappers, combs and brushes, pens, raincoats, computers, bullet-proof vests, and football helmets.

MAKING PLASTIC

Plastic is produced from simple natural ingredients. For example, nylon is made from water, air, and oil. Oil is the main component of many plastics. Rayon is made largely from cellulose—the starch that makes up the cell walls of plants. When the raw materials are combined, chemical reactions cause

◆ The use of more and more plastics has caused their disposal to become a major environmental problem.

atoms to clump together to produce synthetic resins.

Plastic objects are formed in either compression or injection molds. Thermosets are usually formed by compression, or pressure caused by pushing. To make compression plastics, resin powder is placed in a mold. Heat and pressure are then applied to the resin. Thermoplastics are usually formed by injection, or the forcing of a liquid into a mold. In this process, small balls of resin drop into a heated container, where they melt. Then the liquid is forced under pressure into the mold.

CHARACTERISTICS OF PLASTIC

Plastic can be as stiff as steel, as stretchy as a rubber band, or as soft as cotton. It can be used instead of metal, wood, stone, glass, cotton, wool, or ceramics to create lighter, stronger, longer-lasting, rust-resistant, and less costly products. Plastic is used in automobiles because it is lighter than metal. The decreased weight makes cars more energy efficient. Plastic can also be used in medicine to mend damaged bones and to replace heart valves because it is less likely than metal to trigger harmful reactions in the body.

PLASTICS AND THE ENVIRONMENT

As more plastic products are used, disposing of them has become a major environmental problem. It takes a long time for plastic to decay. The uniqueness and artificiality of most plastics make them highly stable and resistant to DECOMPOSITION. After years of being buried in LANDFILLS, plastics remain relatively unchanged.

Some newer plastics are photodegradable. Such plastics break down after extensive exposure to sunlight. However, PHOTODEGRADABLE PLASTICS must remain above ground where they are exposed to sunlight to break down. Even then, they leave a residue behind that must be buried in landfills.

Some plastic can be burned with other GARBAGE. The heat produced may then be used to supply the energy needed to generate ELECTRICITY. However, using plastics in this way requires special devices to remove acid gases and other toxic materials from the smoke given off by burning plastic. These acids, if released into the ENVIRONMENT, cause AIR POLLUTION.

Plastics have contributed to the problems of disposing of SOLID WASTES, increased use of oil (which is used to manufacture plastics), and have had negative impacts on some natural HABITATS.

Recycling is the solution most often recommended for the disposal of plastics. Many communities require citizens to recycle. Different types of plastics are recycled in different ways. For example, thermoplastics are melted and reformed to make new products. Thermosets are shredded or ground into powder for use as insulation. [*See also* BIODEGRADABLE; RECYCLING, REDUCING, REUSING; SCRUBBER; SOLID WASTE INCINERATION; and WASTE REDUCTION.]

Plate Tectonics

❑Scientific theory used to explain CONTINENTAL DRIFT and other dynamic geological processes. Most scientists believe that today's continents were once connected as one large continent called Pangaea a long time ago. Over millions of years, the continents gradually drifted apart and into the positions they occupy today. In plate tectonics, information about Earth's structure and the forces deep within Earth's core is used to explain continental drift.

HISTORICAL DEVELOPMENT OF PLATE TECTONICS

In 1912, German meteorologist Alfred Wegener hypothesized that the continents were once connected in one large landmass. He believed the large landmass broke apart about 200 million years ago. As evidence for this **hypothesis**, Wegener noted the shapes of the continents, which seemed to fit together like pieces of a jigsaw puzzle. He also observed the distribution of certain plant and animal fossils. Wegener argued that this evidence could be explained only by continental drifting, the gradual movement of the continents.

The basic idea of continental drift is widely accepted today. However, at the time Wegener made his proposal, he was criticized by the scientific community. The main criticism was due to Wegener's inability to explain how continental drift occurs.

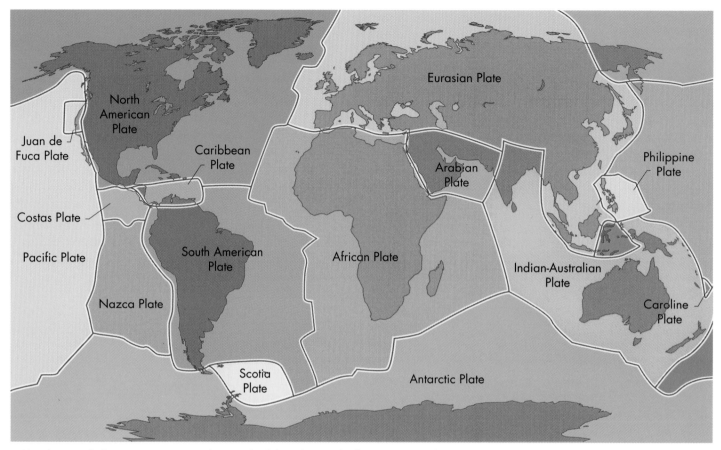

◆ The theory of plate tectonics states that Earth's lithosphere is broken up into eight major plates and seven minor plates.

Acceptance of the continental drift hypothesis began to change in the 1950s and 1960s when scientists began studying Earth's structure in detail. An important piece of information came from Princeton scientist Harry Hess, who proposed that the ocean floors were spreading and moving apart due to strong forces deep within Earth's core. This movement is called *seafloor spreading*. With the discovery of seafloor spreading, scientists began to understand how Earth's crust, or LITHOSPHERE, works.

PRINCIPLES OF PLATE TECTONICS

The theory of plate tectonics is based on information about continental drift and seafloor spreading. According to this theory, Earth's lithosphere is not a solid, unbroken mass. Rather, it is broken up into 15 distinct sections, called *plates*, that float on the semisolid, plasticlike layer of Earth beneath the lithosphere. Connected to some of the plates are the continents, which float like giant rafts in the vastness of Earth's oceans. Scientists have estimated that the continents are currently drifting at a rate of .75 inches (1.9 centimeters) per year. Thus, over millions of years, the continents have moved thousands of miles.

Plate tectonics can be used to explain geological events such as

◆ The San Andreas Fault in California is the boundary between two plates that are prone to sliding past each other.

Plutonium

❚A radioactive substance sometimes used in the production of NUCLEAR POWER. Unlike URANIUM, plutonium is not available in Earth's crust in any significant amounts. Thus, most plutonium is produced artificially.

Plutonium is used in BREEDER REACTORS when uranium undergoes the process of NUCLEAR FISSION. It is also a waste product in nuclear reactors that are fueled by uranium. Plutonium and other radioactive substances are important nuclear FUELS because they produce many times more energy than can be produced by equal amounts of FOSSIL FUELS.

Plutonium was discovered in the late 1930s and early 1940s, during nuclear fission experiments conducted by various scientists. Scientists were interested in the radioactive properties of plutonium and its ability to produce enormous amounts of energy. On July 16,

earthquakes, VOLCANISM, and mountain-building. Various types of interactions at plate boundaries are known to cause these natural events. For instance, when two plates slide past one another, the vibrations caused by this movement create earthquakes. The famous San Andreas Fault in California represents the boundary between the Pacific Plate and the North American Plate. Movements between these two plates caused two of the largest earthquakes in the recent history of the United States—the 1989 earthquake in San Francisco and the earthquake that shook Los Angeles in 1992.

Plate tectonics can also explain the process of mountain formation. Mountain ranges form gradually when two plates collide and crumple up. For example, the Himalaya Mountains in Southeast Asia were formed when the Indian-Australian Plate crashed into the Eurasian Plate.

In areas where two plates move apart, volcanoes are common. As plates move apart, melted rock deep inside Earth is able to reach the surface. Once at the surface, the melted rock is called *lava*. As the lava cools and hardens, it forms new land at Earth's surface. [*See also* CONVERGENT EVOLUTION; EVOLUTION; GEOTHERMAL ENERGY; and NATURAL DISASTERS.]

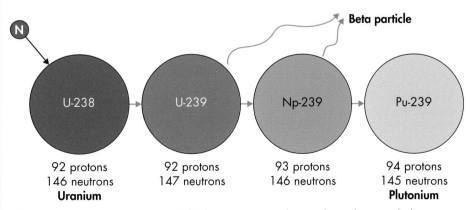

◆ In a nuclear reactor, uranium is hit by neutrons and goes through several changes before it becomes plutonium. Plutonium can be used as a nuclear fuel, or it can be treated as nuclear waste.

1945, plutonium was put to use in the first nuclear explosion in a NUCLEAR WEAPONS test in New Mexico.

In 1951, the first experimental breeder reactor, which converts uranium into plutonium, was built in Idaho. This reactor successfully demonstrated that breeder reactors could be used to extend uranium supplies and also produce energy. However, safely disposing of the nuclear waste produced by nuclear reactors is a serious problem because it stays radioactive for thousands or even millions of years. [*See also* ALTERNATIVE ENERGY SOURCES; ATOMIC ENERGY COMMISSION (AEC); INTERNATIONAL ATOMIC ENERGY AGENCY (IAEA); NUCLEAR WINTER; RADIATION; RADIATION EXPOSURE; RADIOACTIVE FALLOUT; and RADIOACTIVITY.]

Poaching

▮The illegal HUNTING or fishing of organisms. Poaching activities threaten the survival of a variety of animals throughout the world. Such activities can cause the EXTINCTION of SPECIES.

In the United States, poachers slaughter walruses for their tusks, raptors for their feathers, and sturgeon and paddlefish for their eggs, or roe. Some people pay thousands of dollars for the chance to hunt polar bears and GRIZZLY BEARS, mountain lions, jaguars, and other animals that are protected by law. The U.S. FISH AND WILDLIFE SERVICE estimates that illegal profits from poaching in this country alone are about $200 million a year.

The problem of poaching is worse in other parts of the world. In Africa, for example, poachers killed half of all the ELEPHANTS on the continent between 1979 and 1989. Poaching activities also threaten the survival of the endangered mountain GORILLA. One effort to stop poaching worldwide was the enactment of the CONVENTION ON INTERNATIONAL TRADE IN ENDANGERED SPECIES OF WILD FLORA AND FAUNA

◆ Crocodiles and other wildlife are hunted in the Amazon for export to the United States.

◆ Many birds are hunted for their beautiful feathers which are later sold.

◆ The skin and fur of an endangered ocelot are sold in a Panama market.

(CITES). This international treaty contributed to the 1989 ban on ivory sales, which resulted in a significant decline in elephant poaching.

Poaching has also become a serious problem in ocean ENVIRONMENTS. During the 1970s, most nations extended their claims on territorial waters from 12 to 200 nautical miles (22 to 370 kilometers) offshore to gain control of valuable fishing grounds. As a result of this action, fishing grounds that were once open to all became the property of individual nations. If these nations regulate fishing in their waters, many species will thrive. However, it is important to note that unregulated fishing activities could lead to the extinction of some fish species. [*See also* BIODIVERSITY; ENDANGERED SPECIES; ENDANGERED SPECIES ACT; MARINE MAMMAL PROTECTION ACT; NATIONAL WILDLIFE REFUGE; and WILDLIFE CONSERVATION.]

Point Source

▶ A single direct source of WATER POLLUTION, usually from a pipe. Point sources release point POLLUTION—the discharge of pollution from a single source. Point sources are major contributors to the pollution of lakes, rivers, streams, and OCEANS.

Point pollution comes from offshore, as well as from land-based sources. LANDFILLS, underground fuel tanks, and WATER TREATMENT plants are common land-based sources of point pollution. In the oceans, leaky oil tankers, fishing boats, and recreational boats are point sources of pollution.

Point pollution can be controlled and prevented because the source is directly identifiable. Unlike nonpoint pollution, which comes from many sources, point pollution can be observed and therefore controlled. However, even when the source of pollution is known, it is not always easy to eliminate point sources. For example, in the early 1980s, citizens of Canton, North Carolina, asked the local government for strict controls on a paper plant that was dumping dangerous wastes into the nearby Pigeon River. Other citizens disagreed. They were concerned about the jobs that would be lost if the plant closed. In 1985, despite protests from environmental groups, citizens voted to approve a POLLUTION PERMIT for the paper plant. The permit gave the plant the legal right to release pollutants into the river.

Four years later in 1989, environmental groups won the battle against the paper plant. The plant agreed to spend the money needed to buy better equipment to reduce

◆ Polluted mine drainage is being released into a river from a point source.

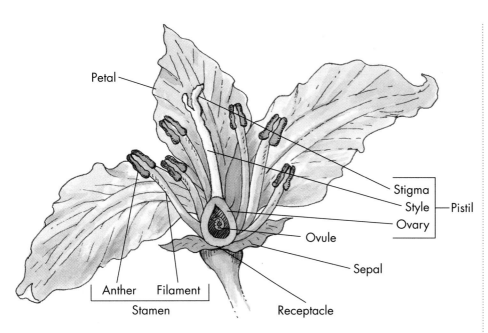

Petal

Anther Filament

Stamen

Stigma
Style — Pistil
Ovary

Ovule

Sepal

Receptacle

◆ The transfer of pollen from stamen to pistil may be done by wind, insects, or other small animals.

There are two kinds of pollination: cross-pollination and self-pollination. In cross-pollination, the pollen from an anther of one flower is transferred to the stigma of another flower. In self-pollination, the pollen from an anther of one flower is transferred to the stigma of the same flower, or to the stigma of another flower on the same plant. In the case of cross-pollination, more GENETIC DIVERSITY is obtained since the offspring have two different parents. In self-pollination, the offspring are genetically identical to their parent.

SELF-POLLINATION

Examples of plants that pollinate themselves are the garden pea and wheat. In the case of the garden pea, pollination occurs even before the flower opens.

Many mechanisms have evolved that decrease the probability of self-pollination—which leads to genetic sameness—and increases the probability of cross-pollination. Even if

pollution. Today, Pigeon River is cleaner than it has been in 80 years. [*See also* EFFLUENT; INDUSTRIAL WASTE TREATMENT; LEACHING; MARINE POLLUTION; NONPOINT SOURCE; OCEAN DUMPING; OIL POLLUTION; OIL SPILLS; SEWAGE TREATMENT PLANT; and THERMAL WATER POLLUTION.]

the *stamen*. The stamen is made up of the anther, which produces pollen, and which is held up by a filament. Flowers may have both male and female structures, or they may have only stamens or only pistils.

Pollination

❙The transfer of pollen from an anther to a stigma in the flower of a FLOWERING PLANT, or from the male cone to the female cone in gymnosperms. In the flower structure, the female part is called the *pistil*. The pistil is made up of a stigma, which is sticky; a style or stalk; and an ovary. The male part is called

◆ Flowers are pollinated by many kinds of insects, such as the dewy dragonfly.

flowers have both stamens and pistils, they may be incapable of self-pollination. Sometimes, the sex organs mature at different times or they may be situated so that self-pollination is physically impossible. In the case of the fuchsia flower, for example, the stigma protrudes beyond the stamen, making it impossible for the pollen from the stamen to reach the stigma.

CROSS-POLLINATION

Cross-pollination in angiosperms, the flowering plants, is done by many different pollinators. More than 20,000 different species of bees are pollinators as well as other INSECTS, such as beetles, wasps, flies,

◆ A butterfly and a crab spider visit a thistle flower.

butterflies, and moths. Bats, BIRDS, snails, slugs and humans are also able to transfer pollen.

The insect-flower relationships began 120 million years ago, when angiosperms evolved. Neither animals nor PLANTS were adapted for pollination. Mutations subsequently occurred which increased a plant's distinctiveness, and mutations in potential pollinators occurred that enabled the animal to recognize the food source. Animals expend a great deal of energy in searching for food. For instance, birds search for food sources by flying. They gather food most economically when they recognize at a distance what plants are best for them. Pollen is very protein-rich, and nectar, produced by petals, is rich in sugar.

Such aspects of the flower as color, size, shape, and fragrance developed so that pollinators are attracted to the same species. In this way, the pollen of one species would not be deposited on the stigma of a different species and be of no use to it.

BEES, BEETLES, AND FLIES

Bees have color vision. They are able to see most colors, except reds, that humans can, as well as the near ultraviolet range, so they can see patterns on some flowers that humans cannot see without special equipment. Flowers pollinated by beetles have a strong or fruity perfume and are usually white or dull-colored.

Some fly pollinators are attracted to foul odors. The stapelias of southern Africa are dull red or

brown and have an odor like rotten meat, which attracts short-tongued flies. The skunk cabbage of the United States emits a foul odor which attracts pollinating flies.

Insect pollinators are critical to people. About one-third of our food plants are pollinated by insects such as bees, wasps, or flies.

BIRDS AND BATS

Birds are very active and fly in search of food. They must eat frequently. They need to recognize food plants at a distance so they do not waste their energy. Hummingbirds have good vision and the flowers they pollinate are bright red or yellow. Bat-pollinated flowers open at night, when bats fly in search of food. These flowers have dull colors.

WIND-POLLINATED FLOWERS AND CONES

Examples of wind-pollinated plants are grasses, oaks, and all cone-bearing trees, such as pines, spruces, and redwoods. Wind-pollinated plants do not have bright colors, fragrances, and nectar because they do not need to attract pollinators. Their flowers tend to be very small, and they frequently have no sepals or petals. Their stigmas are large and are sometimes feathery or sticky, allowing them to catch pollen. Since the chance that pollen will land on the appropriate stigma is small, the anthers of wind pollinated plants produce huge amounts of pollen. The plants grow in dense stands so they are in the right place when the pollen is blown from the anthers or male cones.

Pollution

❚Any potentially harmful substance or form of energy released into the ENVIRONMENT. Pollution is the general term used to describe any substance or form of energy that contaminates air, water, or SOIL. Substances that act as pollutants can enter the environment as solids, liquids, or gases. Forms of energy that may be considered pollution include thermal, or heat, energy; sound energy; and some forms of RADIOACTIVITY. The presence of pollutants in the environment may be harmful to human health, harm organisms other than people, disrupt the beauty of natural environments, or disrupt natural functions or cycles of ECOSYSTEMS.

Much of the pollution in the environment results from human activities, such as the burning of FUELS or the release of heated water from power plants into aquatic ecosystems. However, some forms of pollution result from natural processes including volcanic eruptions; the emission of gases such as METHANE from **swamps**; FOREST FIRES; and a type of deposition called *siltation*.

AIR POLLUTION

AIR POLLUTION results when harmful substances are released into the ATMOSPHERE. Air pollutants are generally classified as gases or PARTICULATES. Particulates are small solids, such as soot and ash, that are suspended in air. Because air is everywhere, air pollution can occur in both outdoor and indoor environments.

Many pollutants in air come from natural sources, including sandstorms, windstorms, volcanic eruptions, forest fires, ocean sprays, and PLANTS. All of these sources of pollution release substances that can break down or harm ABIOTIC FACTORS in the environment, disrupt the natural beauty of the environment, or harm organisms by irritating the respiratory system and eyes or by interfering with respiratory processes. Pollen is an example of a natural pollutant that can be irritating to the eyes and respiratory processes of organisms. The major sources of air pollution that result from human activities are the burning of fuels, such as those used in the operation of factories and fuel-powered vehicles.

NOISE POLLUTION

Sound intensity, or energy, is measured in units called *decibels*. The softest sounds humans can hear have an intensity of about 0 decibels. In contrast, music played at a

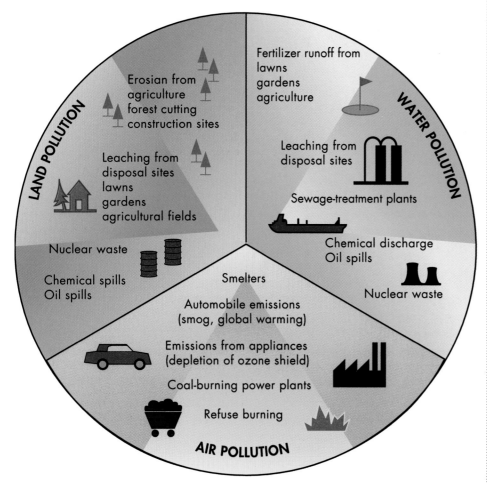

◆ Various ways are being developed and implemented to reduce pollution caused by human activities.

rock concert often has an intensity greater than 100 decibels. Loud or high-pitched sounds, with an intensity greater than 120 decibels, can be harmful to living things. The presence of such sounds in the environment is called NOISE POLLUTION.

Noise pollution may be only an annoyance. However, it can also cause stress-related disorders and a loss of hearing. For example, exposure to loud noises over a period of time can place stresses on the body that lead to an increase in blood pressure, digestive disorders, and emotional or psychological problems. Common sources of noise pollution include airplanes, automobile horns, automobile engines, lawn mowers and other yard equipment, machines used in factories, and the playing of loud music on stereo systems or at concerts. To avoid disorders that may be caused by extended exposure to such noises, many people whose work or home environments contain such noises wear earplugs or ear protection.

WATER POLLUTION

WATER POLLUTION results when harmful substances or energy are released into freshwater and saltwater environments at Earth's surface. Water pollution also affects water beneath the surface, such as that in groundwater supplies and AQUIFERS. Some water pollution results from natural processes including DECOMPOSITION. Most water pollution, however, results from human activities, such as the intentional dumping of wastes from factories and the accidental release of harmful chemicals from OIL SPILLS. Water pollution

also results from the LEACHING of toxic materials from mine TAILINGS and LANDFILLS and from poor LAND USE and farming practices.

THERMAL POLLUTION

Thermal pollution results from the release of heat into the environment. Most often, thermal pollution is caused by power plants (both nuclear and coal burning) or other types of industrial factories. Thermal pollution usually affects aquatic environments, such as shallow bays, lakes, and rivers.

In many factories and power plants, water from rivers, lakes, or bays is used in a cooling system to reduce heat given off by industrial processes. This water acts as a **coolant** by absorbing the heat given off by various devices and processes inside the plant. This heated water may then be returned to the river, lake, or bay from which it was collected. When the heated water is returned to the body of water, it raises water temperatures within the ecosystem.

A rise in water temperature can harm organisms living in the water. Thus, the release of such energy is a type of pollution known as THERMAL WATER POLLUTION. Many water organisms, such as FISH and AMPHIBIANS, have body temperatures that are regulated by the temperatures of their surroundings. As temperatures in the water increase, so does the body temperature of the organism. This, in turn, causes many functions of the organisms, such as respiratory rate and heart rate, to speed up. The problem is worsened by the fact that increasing the temperature of water may alter

the DISSOLVED OXYGEN content of the water. Cooler water can hold more dissolved oxygen than warmer water. Thus, at the same time an organism's respiratory rate increases, the OXYGEN available to the organism decreases. As a result, the organism may suffocate and die.

LAND POLLUTION

Land pollution occurs when substances that are potentially harmful

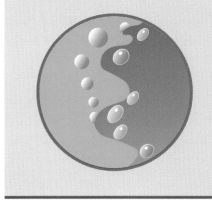

to the environment or its organisms are released into soil or permitted to accumulate on land. Pollutants that affect soils may be solids, liquids, or gases. The release of such substances into soil can make the soil unsuitable for the many types of organisms that live there.

One of the major sources of land pollution are SOLID WASTES. Such wastes include all GARBAGE, refuse, and SLUDGE that are produced by communities, as well as those resulting from agriculture, MINING, and FORESTRY.

At one time, people chose a location in which to dispose of solid wastes and allowed the wastes to accumulate in large piles upon the land. However, the unsanitary conditions resulting from this practice, along with the foul odors and ugly appearance of the discarded wastes, led most areas to develop laws forbidding this practice. As an alternative, wastes began to be buried beneath the soil's surface in areas called landfills.

Landfills, especially after they were filled in and closed to additional waste disposal use, solved much of the odor and appearance problems resulting from open dumping. However, it was soon discovered that toxic substances that accumulated in rainwater washing through a landfill were often carried away in the rainwater to nearby bodies of water or into the groundwater supplies people used for drinking water. To solve this problem, the federal government created strict rules for how landfills could operate and required all landfills to be lined with clay and plastic to allay, or at least help reduce, the chances of harmful substances leaching out of the landfill.

Soil is also polluted from such practices as mining and agriculture. All of these practices may cause immediate problems in the environment by destroying HABITATS. Many people also believe that the immediate changes in the environment

resulting from such practices disrupts the natural beauty of the environment. These practices directly release substances into the soil that may be harmful to organisms living there. In addition, because many substances are easily dissolved in and carried away by water, RUNOFF and LEACHING may carry pollutants present in soil into groundwater supplies and aquatic environments.

RADIOACTIVITY

Substances that are radioactive release particles and energy into the environment. These particles and energy are called RADIATION. Radiation can be harmful to organisms because some forms of radiation easily penetrate the cells and tissues of organisms. Once taken into the cells and tissues, the energy in the

---

Potential Benefits of Reducing and Preventing Pollution

1. Improved human health and decrease in worker stress
2. Increased real estate values
3. Higher profits from improved agricultural and forest production and commercial marine harvests
4. Improved recreational uses, benefiting associated businesses
5. Extended lifetime and reduced maintenance of materials

Pros and Cons of the Different Methods of Improving the Environment

Method	Pros	Cons
Prohibition	Prevents or eliminates the damage	Very expensive
Direct regulation	Controls pollution and keeps it to a minimum	Difficult to enforce and identify the many sources of pollution
Incentives	Makes it profitable for polluters to reduce pollution or not to pollute at all	May encourage polluters to collect payments later
Suing for damages	Allows compensation of individuals or groups	Time-consuming, expensive, and difficult to establish who damaged whom and to what degree
Appeal to the public's moral sense	Increases awareness and sensitivity to issues	Often results in guilt and discontent instead of action

radiation may actually destroy cells and tissues or alter the DNA within the cells of the organism. Some forms of radiation have also been linked in humans to such diseases as CANCER.

Some radioactive materials are used by people for medical treatments and diagnoses, including those involving X RAYS. Because of the huge amount of energy they give off, some radioactive materials, such as URANIUM, are used in nuclear power plants to produce ELECTRICITY.

When radioactive materials for use by people are mined, RADIOACTIVE WASTES are often released into the environment where they can disrupt the life processes of organisms. Other radioactive wastes result from radioactive materials used as fuels in nuclear power plants. Even after they are used to produce energy, nuclear fuels continue to be radioactive—sometimes for millions or billions of years. Disposal of these wastes, which are no longer useful to people, is a problem because radiation is able to pass through many solid substances. Concerns about how to safely dispose of such materials have led many communities to reject the building of new nuclear power plants. [*See also* ACID RAIN; AGRICULTURAL POLLUTION; AGROECOLOGY; AIR POLLUTION CONTROL ACT; ALGAL BLOOM; CLEAN AIR ACT; CLEAN WATER ACT; DETERGENTS; UNITED NATIONS EARTH SUMMIT; ENVIRONMENTAL PROTECTION AGENCY (EPA); EUTROPHICATION; GLOBAL 2000 REPORT, THE; GLOBAL WARMING; GREENHOUSE EFFECT; HAZARDOUS SUBSTANCES ACT; INDUSTRIAL WASTE TREATMENT; INTERNATIONAL TRADE IN TOXIC WASTE; MARINE POLLUTION; MONTREAL PROTOCOL; OCEAN DUMPING; OIL POLLUTION; OZONE POLLUTION; POLLUTION PERMITS; SEWAGE; SEWAGE TREATMENT PLANT; SICK BUILDING SYNDROME; SOLID WASTE DISPOSAL ACT; THERMAL WATER POLLUTION; TOXIC SUBSTANCES CONTROL ACT (1976); WASTE REDUCTION; and WATER QUALITY STANDARDS.]

Pollution Permits

▶ Permits issued by state and federal agencies that allow the discharge of pollutants into the ENVIRONMENT. Under the law, factories, power plants, and other industrial sites must obtain pollution permits from government authorities such as the ENVIRONMENTAL PROTECTION AGENCY (EPA) before they can operate and release pollutants.

The United States has some of the toughest environmental laws of any industrialized nation. These laws set safe levels of different types of pollutants. Scientific studies do not always determine what a safe level is for a type of pollutant. Thus, the standards for what is considered safe sometimes lead to changes in the law.

Under the CLEAN AIR ACT, all stationary industrialized sites, such as factories, must obtain pollution permits that set strict limits on emissions. In addition, all stationary sources of AIR POLLUTION must use the BEST AVAILABLE CONTROL TECHNOLOGY (BACT) to manage these pollutants, regardless of cost.

Pollution permits help protect ECOSYSTEMS from the damaging effects of many pollutants. They also serve to protect human health by making the public aware of the types and amounts of pollutants released by industry. [*See also* AIR POLLUTION; EFFLUENT; POINT SOURCE; PRIMARY POLLUTION; SOLID WASTE DISPOSAL ACT; THERMAL WATER POLLUTION; WATER POLLUTION; and WATER QUALITY STANDARDS.]

Polychlorinated Biphenyls
See PCBS

Population Growth

▶ An increase in the number of individuals of a given SPECIES. Every organism depends on Earth's resources for its survival. For example, all organisms need food, space, water. These resources are naturally limited, for Earth is a finite place. Most species produce more offspring than can survive, given the limited resources available. In nature, limited resources maintain fairly stable species populations. The "extra" individuals die due to predation, lack of resources, and other factors.

If unchecked, a species' population growth will quickly increase beyond the capacity of that species' HABITAT to support it. The resource limit any ECOSYSTEM has for support of a given species is called the CARRYING CAPACITY. For example, if limited supplies of food and water

♦ As the human population grows, many natural areas are used as living space.

have learned how to manipulate nature to make it produce those resources we need to survive. For example, the first dramatic increase in the human population occurred thousands of years ago during the AGRICULTURAL REVOLUTION. The increased food supply enabled many more people to survive. The population size increased again during the INDUSTRIAL REVOLUTION. Today, technologies of all kinds—especially medical advances—continue to improve the chances (and duration) of survival for nearly every person on Earth.

The development and growth of agriculture transformed a "natural" habitat filled with wild creatures into a controlled habitat dedicated to the production of food for people. In effect, agriculture has increased Earth's carrying capacity for human beings. Even today, we continue to take over natural areas to increase food resources for ourselves.

Expanding our resource base into natural areas is harming BIODIVERSITY. When people take over wild lands and plow them for agriculture, native PLANTS and animals are forced to leave to find other suitable habitats. Some organisms may die once their habitat is destroyed.

did not check the world's population of ELEPHANTS, their numbers would increase from about 1 million to about 20 million in a mere 200 years. However, the carrying capacity of elephant habitat (the finite amount of food and water) limits the number of elephants that can survive there. In addition, elephants reproduce very slowly.

If predation did not control INSECTS such as mosquitoes—organisms that produce tens of thousands of offspring each year—the world would soon be buried by these insects. In time, however, the mosquitoes would perish—their populations would overwhelm their habitat's carrying capacity and crash—for there are not enough resources to maintain such huge populations of these animals.

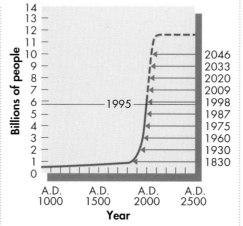

♦ Exponential growth results in a population explosion.

AGRICULTURE AND HABITAT DESTRUCTION

Humans are especially susceptible to the problems of excessive population growth. We are clever. We

KINDS OF POPULATION GROWTH

The human population is growing *geometrically*: each generation is growing as a multiple of the previous one. For example, today there may be 5 billion people on Earth; in 20 years there are expected to be 10 billion people; in 40 years there may be 20 billion. Notice that each increase represents a multiple of two; in every generation the population doubles.

Food production, on the other hand, increases *arithmetically*: a given amount is added each year. For example, today a field may produce 1,000 bushels of wheat; next year it may produce 1,100 bushels; the year after that it may produce 1,200. Through improved agricultural methods, each year the field increases its yield by 100 bushels of wheat: 100 bushels are added each year. However, there is no way an arithmetically increasing food supply can keep pace with a geometrically increasing population. At some point, the population will crash due to lack of food.

The world is facing a collision between human populations and resources. In developed nations, population growth is low but resource use is high. (The United States has 6% of the world's population and uses nearly 50% of its resources.) In developing countries, exponential population growth has led to overexploitation of the land for food and FUEL. DEFORESTATION and resulting DESERTIFICATION are ruining previously rich habitats. Biodiversity is in decline in areas with high human population growth. The degraded ENVIRONMENT makes survival even harder for the exploding human population. [*See also* DEMOGRAPHY; ENVIRONMENTAL EDUCATION; and SPECIES DIVERSITY.]

Porpoises

See DOLPHINS/PORPOISES

Prairie

▶ Name given to the North American GRASSLANDS. Similar broad stretches of flat land are found throughout the world's TEMPERATE ZONES, commonly lying 300 to 1,500 feet (90 to 450 meters) above sea level. They are known as *steppes* in Central Asia; *pampa, campo,* and *llano* in different parts of South America; and *veldt* in Africa. All, however, are BIOMES in which grasses are the dominant vegetation. Grasses depend on rainfall during growing seasons, because their shallow roots cannot reach groundwater deep below the surface. Thus, PRECIPITATION determines

whether a level region becomes grassland. Annual rainfall of 10 to 14 inches (250 to 1000 centimeters) can sustain a grassland. Rainfall less than this will turn the region into a DESERT; more rainfall will turn it into a FOREST. Yet grassland CLIMATES are typically unpredictable. Because PRECIPITATION may vary wildly from month to month and year to year, droughts and grass fires are common. Fire plays a major role in maintaining grasslands, burning off trees and other foliage at the surface without harming the extensive root networks of the grass. Grasslands are tremendously fertile—a hectare of grassland, as one British naturalist has observed, can support a greater weight of living flesh than any other kind of country. For example, before the arrival of Euro-

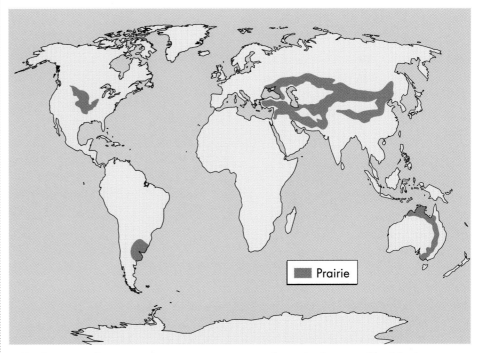

◆ The location of prairies in temperate zones are shown in the map. The North American prairies provide an ideal place for growing grains for human food. Grains are the seeds of grass.

peans, the prairie was home to bison numbering in the tens of millions. Grazing animals also play a major part in maintaining grasslands, as does fire, both of which tend to favor grasses and other herbaceous PLANTS, and discourage shrubs and trees.

Grasses were evidently forest-glade plants when they first appeared about 65 million years ago. The mountain-building that resulted from continental movements reduced the amount of moisture that reached the interiors of some landmasses. In response, forests receded and grasslands expanded. Hoofed animals coevolved on these changing landscapes. Human beings themselves originated on the grassy savannas of eastern and southern Africa. Millions of years later, agricultural societies developed in the grasslands of the Middle East.

Europeans transferred Old World agricultural practices to America in the sixteenth century, but these practices proved inadequate on the prairie. The rich prairie sod was too tough and heavy to be turned by European-style wooden plows.

Also, annual precipitation was below the minimum needed to support nonirrigated agriculture.

During the nineteenth century, a succession of technological innovations made exploitation of the prairie possible. The all-steel plow and a practical well-drilling method were perfected in the 1830s. The self-regulating wind pump, invented in 1854, could be coupled with a drilled well to tap the WATER TABLE deep beneath the plains. Following the introduction of barbed wire in 1873, LIVESTOCK could be effectively confined on the open grassland. Prairie farming became increasingly dependent on heavy machinery, and the growing demand for reapers, binders, riding plows, gang plows, and multiple-row cultivators was a major force in

the industrialization of the United States.

The creation of the North American grain belt was even furthered by several decades of above-average rainfall, which encouraged the cultivation of areas that normally could not support agriculture. However, a period of extended drought followed on the Great Plains that caused the economic devastation of the DUST BOWL during the 1930s. Human influence on the prairie has not been beneficial. Most grasslands have overgrazed. The nutrients in the SOILS have been depleted by wind and water EROSION. Water tables have become depleted, too. Earth's fertile natural grasslands once covered almost one-third of the terrestrial land surface, but this is no longer true. The largest

◆ Prairies are critical biomes, since they provide the grains, or grass seeds, of the human food supply.

◆ Prairie dogs dig tunnels for safety.

expanse of surviving natural prairie in North America is in the Flint Hills of Kansas. Other tracts are preserved in YELLOWSTONE NATIONAL PARK and in preserves such as the NATIONAL GRASSLANDS and in prairie preserves owned by conservation groups such as the Nature Conservancy. Often, management of such preserves involves PRESCRIBED BURN, the elimination of EXOTIC SPECIES of plants, and the reintroduction of NATIVE SPECIES. [*See also* BUREAU OF LAND MANAGEMENT; COEVOLUTION; and CONTINENTAL DRIFT.]

Precipitation

▐▶Forms of water falling to Earth from the ATMOSPHERE, including rain, snow, hail, and sleet. Precipitation begins in the CLOUDS as small droplets. In very cold clouds, ice crystals form around dust particles. The tiny cloud droplets collect on the ice crystals. It takes several million cloud droplets to produce a precipitation particle heavy enough to fall from the cloud. As it falls, the particle collects more droplets. Depending on the temperature of the air it falls through, the precipitation particle may hit the ground as snow, hail, or sleet.

When clouds are not cold, particles of chemicals from salt spray or sulfur substances in the atmosphere are the nuclei for the formation of precipitation. More and more cloud droplets collect on the chemical particle. It begins to fall, collecting more cloud droplets as it falls, becoming heavy precipitation that strikes Earth as rain.

Precipitation is a valuable source of fresh water. It plays an important role in Earth's WATER CYCLE. In this cycle, water vapor given off by Earth's OCEANS, lakes, PLANTS, and animals drifts upward into the atmosphere. As the vapor cools, it again becomes liquid. Precipitation returns the water to Earth as fresh water. [*See also* ACID RAIN; BIOME; CLIMATE CHANGE; DESERT; EVAPOTRANSPIRATION; NATURAL DISASTERS; and WEATHER.]

◆ Roots of Siberian irises take in water that runs into the soil.

◆ Precipitation falls to Earth, providing water for plants and animals.

Predator

▐▶Any organism that kills and eats another organism. The organism killed by a predator is known as the *prey*. The relationship between predator and prey is called *predation*.

Predators most often include animals that eat other animals. However, a few PLANTS that can eat animals, such as the Venus's-flytrap and the pitcher plant, are also predators. Predatory animals are often large animals, such as wolves, lions, and toothed WHALES. However, many small animals, such as many INSECTS, spiders, and salamanders, are also predators.

Many large predators, such as TIGERS, mountain lions, and wolves,

♦ A lion family feeds on a zebra.

♦ A timber wolf feeds on a dead moose.

♦ The American alligator is a predator commonly found in swamps in North Carolina and Florida.

example, many large predators are hunted for their skins or pelts, as trophies, or in some cases for certain internal organs. For instance, tigers are sometimes hunted for their claws, which are used as ashtrays. GRIZZLY BEARS are often hunted for their head or for their pelts, which are used as rugs.

Many large predators, such as wolves, are considered beautiful by some. These animals have become popular symbols for wildlife protection. Large predators such as snow leopards and lions are also popular zoo animals. [*See also* CARNIVORE; CONSUMER; ENERGY PYRAMID; and TROPHIC LEVEL.]

Prescribed Burn

▶A fire started and controlled by people to maintain the health of ECOSYSTEMS. In some cases, fire is needed for the growth of some organisms in the ecosystem. In others, the fire helps to clear the ecosystem of debris.

Fire can be a destructive force in natural ENVIRONMENTS. However, in some ecosystems, fire is needed to allow certain SPECIES to thrive and to keep the ecosystem balanced. Some coniferous trees, for example, require the heat produced by fire to release the seeds contained in some of their cones. The jack pine and lodgepole pine are two such trees. One of the best examples of an ecosystem in which periodic fires help the ecosystem thrive is that of the GRASSLAND OR PRAIRIE.

are considered threatened or ENDANGERED SPECIES. There are several reasons for the decline of large predators. One reason for the decline results from the need of many large predators for large areas for HUNTING. HABITAT LOSS is especially deadly to such predators. In addition, toxins that enter FOOD CHAINS and FOOD WEBS may become concentrated in predators through the process of BIOACCUMULATION. For example, many panthers in the EVERGLADES NATIONAL PARK suffer from poisoning caused by MERCURY that has been passed through the food chain.

Overhunting by humans and illegal POACHING also threaten the survival of some predators. For

Grasses use most of their energy to produce roots. Natural fires, most often caused by lightning, burn off the tops of the grasses, while leaving the roots intact. Following the fire, these grasses produce new shoots in great abundance. Grasses regenerate after a fire far more quickly than do trees or bushes. Thus, fire helps grasses dominate prairie ecosystems. Fire also clears the ground of dead PLANTS and organic debris, making the nutrients contained in these materials available to the SOIL, where they can be taken in by plant roots. In this way, fire helps to enrich soil.

Most of North America's native prairie has been destroyed by farming. This destruction has occurred mostly as EXOTIC SPECIES grown as crops replace the native vegetation of the area. Some few remaining tracts of native prairie are protected by federal or state governments or by private conservation organizations, such as the Nature Conservancy. To maintain the health of grassland ecosystems while still protecting the property and lives of people who live on and farm the region, scientists and land managers have developed a system of prescribed burning.

Several basic conditions and techniques are involved in a successful prescribed burn. The prescribed burn must occur on a day with little or no wind so that the fire is not blown out of control. To prevent an uncontrolled spread of fire, drought conditions, in which vegetation is tinderbox dry, cannot be present. Before the burn begins, a firebreak is created around the area to be burned. A *firebreak* is a patch

◆ Preparation is being made for a controlled forest burn.

◆ A prescribed burn is in progress.

◆ A firebreak is used as a safeguard to prevent the spread of fire.

of land cleared of vegetation and anything else that might burn. The firebreak is used as a safeguard to prevent the fire from escaping the burn area. Other safety precautions, such as a readily available water supply, are also put in place. Once the fire is started, a team of prescribed burn experts monitors it constantly. In most cases, all goes well and the fire does its job in its restricted area.

Prescribed burns are not limited to grasslands. They are also used to promote SUCCESSION in some FORESTS or to control unwanted species of shrubs in forests. The methods used for a prescribed burn of a forest are much the same as those used in grasslands. In fact, any natural area

that requires fire to maintain the vitality of the ecosystem can make use of prescribed burn methods. [*See also* ADAPTATION; CONIFEROUS FOREST; FIRE ECOLOGY; FOREST FIRE; NATIONAL GRASSLAND; NATIONAL PARK SERVICE; NATURAL DISASTERS; NITROGEN CYCLE; and YELLOWSTONE NATIONAL PARK.]

Primary Pollution

▶Pollutants that are released, usually through human activities directly into the ENVIRONMENT. Pri-

mary pollutants are generated by a variety of human activities. Motor vehicles, factories, and power plants discharge many polluting chemicals into the atmosphere and cause AIR POLLUTION. If PESTICIDES and fertilizers used in the agricultural industry wash off the land and enter water sources, they can cause WATER POLLUTION. When humans litter and discard other wastes, they contribute to land POLLUTION. All the above provide examples of primary pollution.

Environmentalists are concerned with all types of primary pollution because of the potential effects on the health of ECOSYSTEMS. Primary pollution can also harm human health.

Humans are not the only sources of primary pollution. Pollution may also originate from natural sources. Active volcanoes, for instance, can release volcanic ash into

◆ A guillemot is a victim of the *Exxon Valdez* oil spill in Alaska.

◆ A volcano is a producer of primary pollution.

◆ A massive amount of air pollution is being produced by factories.

the ATMOSPHERE. The ash is a PARTIC-ULATE which can bring about a CLI-MATE CHANGE. Large amounts of atmospheric ash can also cause ecosystem damage by accumulating on PLANTS and animals and disrupting their abilities to carry out life processes. Other natural sources of primary pollution include bogs and marshes, which release METHANE into the atmosphere. Methane is a GREENHOUSE GAS that is present in NATURAL GAS.

Primary pollutants can cause immediate ecological damage or damage over a period of time. Primary pollutants can also react with other substances in the environment to produce secondary pollutants. NITROGEN OXIDES, for instance, released from AUTOMOBILES and power plants, can react with other primary pollutants, like HYDROCARBONS, SULFUR DIOXIDE, and PARTICULATES, to produce SMOG. [See also CHEMICAL SPILLS; DDT; and EUTROPHICATION.]

Producer

▮An organism that makes its own food using raw materials and energy from its ENVIRONMENT. Except for a few SPECIES of BACTERIA, producers such as PLANTS, ALGAE, and some other protists, make food through the process of PHOTOSYNTHESIS. Producers are commonly referred to as AUTOTROPHS, a term which means "self-feeders."

In most ECOSYSTEMS, producers capture the energy in sunlight for the purpose of making food. This process, known as PHOTOSYNTHESIS, requires sunlight, water, and CARBON DIOXIDE, which the producers obtain from their environment. Through photosynthesis, producers make glucose—the energy-rich sugar that most organisms, including the producers themselves, need to survive. Because producers make energy available for CONSUMERS, almost all life depends on the ecological work of producers. [See also ENERGY PYRAMID; FOOD CHAIN; and FOOD WEB.]

	Common Primary Pollutants and Their Sources	
	Pollutants	**Sources**
Air	hydrocarbons	motor vehicles
	nitrogen oxides	motor vehicles; industry
	sulfur oxides	power plants; industry
	CFCs	air conditioners; refrigerators
Water	heavy metals	industry
	pesticides	homes; agriculture
	fertilizers	agriculture; homes
	disease-causing organisms	raw sewage
Land	garbage	homes; businesses; industry
	toxic wastes	industry; homes
	radioactive wastes	nuclear power plants

Public Land

▶ LAND that is owned by all people of an area, such as a city or country, and managed by a government agency. In the 1870s, the U.S. government realized that some land areas must be preserved. The acquisition of land for the public began when President Ulysses S. Grant signed an act that designated a portion of the state of Wyoming as YELLOWSTONE NATIONAL PARK. At about the same time, other countries began similar practices. As a result, public lands were acquired in all parts of the world and set aside for a variety of uses.

Currently, about 40% of all lands in the United States are owned by the public and managed by federal, state, or local government agencies. More than one-third of these lands are managed by the federal government. Most of this land is located west of the Mississippi River.

Lands managed by the federal government fall into six categories. These categories include the National Wilderness Preservation System, NATIONAL PARKS and monuments, NATIONAL WILDLIFE REFUGES, national resource lands, NATIONAL FORESTS, and military installations.

NATIONAL WILDERNESS PRESERVATION SYSTEM

The National Wilderness Preservation System of the United States occupies about 46,000 square miles (119,000 square kilometers) of land area. In 1964, the WILDERNESS ACT gave the federal government the power to protect undeveloped public land as part of a National Wilderness Preservation System.

In the late 1970s, President Jimmy Carter added millions of acres of WILDERNESS area to this system. Much of this land was located

◆ The Grand Canyon in Arizona provides a sense of history, countless scenic views, and evolutionary fossil records within its walls.

in Alaska. Wilderness areas of the United States are open to hiking, fishing, boating (without motors), and camping. In such areas, the building of roads, timber harvesting, GRAZING, MINING, and commercial activities are forbidden. In addition, no structures may be placed on the land except those that were present at the time the area was designated as a WILDERNESS. Wilderness areas are safe havens for many SPECIES of WILDLIFE.

NATIONAL PARKS AND MONUMENTS

About 125,000 square miles (324,000 square kilometers) of U.S. lands are designated for use as national parks or monuments. The NATIONAL PARK SERVICE manages more than 250 units of national park and monument land. These areas include national parks such as YOSEMITE NATIONAL PARK and Yellowstone National Park, as well as national recreation areas, our national monuments, historic battlefields, trails, rivers, seashores, and cultural sites of national importance.

The goal of the National Park Service is to preserve and protect natural resources, including landscapes and preserves and cultural resources that detail the country's cultural heritage for future generations. It also protects wildlife HABITATS, including the wilderness areas that lie within many national parks. To help reduce the risk of HABITAT LOSS caused by human activities, national parks can be used only for camping, hiking, boating, and fishing. Motor vehicles are permitted only on roads.

◆ Twin Lakes, in the Sierra Nevada Mountains of California, provides many scenic views.

◆ Many skiers go to the Sundance ski area in Utah.

◆ Public lands are put to many uses, such as duck hunting and sight-seeing.

◆ The U.S. Air Force Museum is built on public land.

NATIONAL WILDLIFE REFUGES

The national wildlife refuges of the United States occupy about 140,000 square miles (363,000 square kilometers) of land. The FISH AND WILDLIFE SERVICE manages 437 wildlife refuges. About 25% of the land occupied by wildlife refuges is also designated as wilderness.

Most refuges are designed to protect breeding habitat for waterfowl and big-game animals in order to provide sustainable populations for HUNTING. A few refuges are specifically designed to protect one or more ENDANGERED SPECIES. Activities such as hunting, trapping, fishing, oil and gas development, and livestock GRAZING are permitted in some wildlife refuges. However, these activities are limited to those areas in which the secretary of the interior finds such uses compatible with the purpose of the refuge.

NATIONAL RESOURCE LANDS

National resource lands occupy 460,000 square miles (1.2 million square kilometers) of land in the United States. These lands are mostly scrubland, GRASSLAND, and DESERT located in the western United States or Alaska. National resource lands are managed by the BUREAU OF LAND MANAGEMENT for MULTIPLE USE. Much of the land is used for livestock grazing. In addition, much of the land has been acquired because it contains MINERALS and FOSSIL FUELS that are important to the American economy.

NATIONAL FORESTS

National forests occupy 300,000 square miles (777,000 square kilometers) of U.S. land. The national forests include 20 grassland regions and 156 FORESTS that are managed by the U.S. FOREST SERVICE. Approximately 15% of this land has been established as wilderness. The remaining 85% is managed for sustained yield and multiple use. National forest land is used for timber, grazing, agriculture, hunting, fishing, mining, and oil and gas leasing. In addition, some of the land is leased to businesses for such use as ski resorts.

MILITARY INSTALLATIONS

Military installations occupy 40,000 square miles (100,000 square kilometers) of land. Many military bases occupy important environmental habitats. For example, the Wright-Patterson Air Force Base in Ohio contains the largest tall-grass PRAIRIE in the state. Some military bases are located on habitat for endangered species. The bases are used for such purposes as firing ranges, troop maneuvers, and as dumping grounds for old trucks and tanks the military no longer uses. The bases have not been forested or otherwise disturbed.

In 1990, Congress allotted money for the Legacy Resource Management Program. This program is designed to promote the identification and management of biological resources located on lands used by the Department of Defense. Using this money, Fort Bragg in North Carolina has created a position for an Endangered Species Officer. So far, 1,200 rare plant populations have been identified on this base. It is likely that other military lands will also discover rare species whose habitat must be preserved. [*See also* ARCTIC NATIONAL WILDLIFE REFUGE (ANWR) BUREAU OF RECLAMATION; LAND STEWARDSHIP; MUIR, JOHN; PINCHOT, GIFFORD; RANGELAND; and ROOSEVELT, THEODORE.]

Public Trust

▌A doctrine, rooted in Roman law, by which ownership of lands lying under navigable waters is said to belong to the state, which holds them in trust for the benefit of the public. The doctrine was originally intended to safeguard navigation by protecting waterways from being altered. The term came to have broader meaning as it was applied to many different resources. During the late 1970s, it was given an additional twist by a group of students who had banded together as the Mono Lake Committee. The committee brought legal action against the City of Los Angeles, claiming that destruction of the integrity of an ECOSYSTEM constituted a violation of the old law.

The ecosystem in question was Mono Lake, located in east-central California. The 63-square-mile (162-square-kilometer) salt lake has no outlet. In the past, its water, which supported great flocks of aquatic birds and many other kinds of WILDLIFE, had been kept diluted by fresh water flowing from the Sierra Nevadas. During the 1940s, everthirsty Los Angeles had began to divert much of the fresh water inflow from the mountains, and Mono Lake had began to shrink. As it did, the concentration of salt in its water became greater, as did the prospect of ecological collapse in the area.

Ultimately, the Mono Lake Committee's campaign to save the lake led to a favorable ruling by the State Water Resources Control Board. Under the ruling, Los Angeles would have to reduce its diversions of water sufficiently to allow the lake to recover. Because the ruling allows Los Angeles to take variable and increasing amounts of water as the lake level rises, experts estimate that recovery will take about a quarter of a century. Nevertheless, the compromise averted a major ecological disaster, and the Mono Lake Committee's ingenious use of the Public Trust doctrine has provided environmental activism with a valuable courtroom tool. [*See also* ENVIRONMENTAL ETHICS; SALINITY; and SALINIZATION.]

R

Radiation

▶The transfer of heat from one place to another. Radiation also refers to waves and particles given off by radioactive materials such as URANIUM and PLUTONIUM. Common forms of radiation are light, sound, heat, and the particles emitted by radioactive substances. All forms of radiation involve energy changes. For instance, when a light bulb radiates light, the filament in the bulb has changed the energy of an electric current into light energy and heat energy.

The most abundant source of energy on Earth is the electromagnetic radiation given off by the sun. This radiation powers PHOTOSYNTHE-SIS and is the basis for most life on Earth. Visible light, the light you can see, is made up of many colors. You can observe these colors when sunlight passes through airborne water droplets that form a rainbow.

Electromagnetic radiation travels in waves that have a range of different **wavelengths**. Electromagnetic radiation forms an ELECTRO-MAGNETIC SPECTRUM from gamma rays and X RAYS having very short wavelengths through visible light and microwaves to radio waves having very long wavelengths. Visible light extends from violet light, with a wavelength of about 500 **nanometers** (nm), to red light that has a wavelength of about 750 nm. This is the range of light that PLANTS and other PRODUCERS use for photosynthesis. [*See also* ABIOTIC FACTORS; ELECTROMAGNETIC SPECTRUM; and RADIOACTIVITY.]

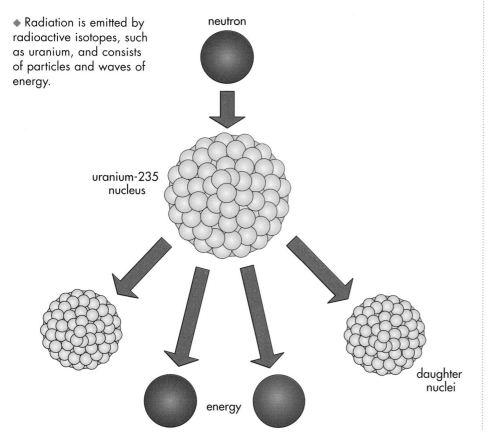

◆ Radiation is emitted by radioactive isotopes, such as uranium, and consists of particles and waves of energy.

neutron

uranium-235 nucleus

daughter nuclei

energy

Radiation Exposure

▶Contact with large doses of RADI-ATION that can cause health problems. When people talk of radiation exposure, they usually mean exposure to radioactivity that can damage biological tissues. One such source of damaging radiation results from the decay of radioac-

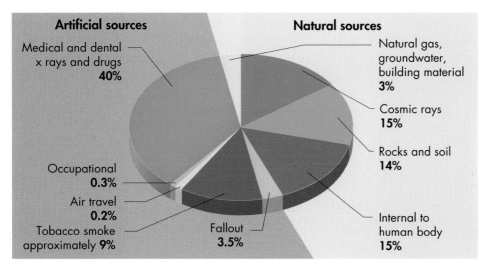

Artificial sources

Medical and dental x rays and drugs **40%**

Occupational **0.3%**

Air travel **0.2%**

Tobacco smoke approximately **9%**

Fallout **3.5%**

Natural sources

Natural gas, groundwater, building material **3%**

Cosmic rays **15%**

Rocks and soil **14%**

Internal to human body **15%**

◆ Humans are exposed to natural sources of radiation, such as cosmic rays, and artificial sources, such as x rays.

tive elements such as URANIUM, radium, and polonium. These elements occur naturally within Earth. They are also used as FUELS in NUCLEAR POWER plants and in NUCLEAR WEAPONS. Exposure to intense radio-activity can result in radiation sickness, which often is fatal. Nuclear radiation is measured in units called *rems*. At a 100 rem exposure, the body's blood-forming tissues are damaged. Exposures greater than 500 rems usually result in death within a few days or weeks. Radiation exposure can also cause birth defects, even years after the exposure occurs.

Damaging radiation is also produced in the form of high-energy wavelengths of electromagnetic radiation. This radiation is given off by the sun as gamma rays, X RAYS, and ultraviolet rays. All these forms of radiation have enough energy to penetrate biological tissues. Once in the tissues, they break apart and destroy many molecules that exist in organisms. Exposure to radiation can be harmful to people, causing problems ranging from a slight sunburn to skin CANCER or even death.

Humans and other organisms are constantly exposed to radiation from sources such as the sun, cosmic rays, and radioactive elements contained in rocks, SOIL, air, water, and food. In addition, medical x rays, wastes from nuclear power plants and weapons manufacture, and radioactive elements used for medical purposes also release radiation into the ENVIRONMENT. [*See also* CHERNOBYL ACCIDENT; NUCLEAR WINTER; PLUTONIUM; POLLUTION; RADIOACTIVE FALLOUT; RADIOACTIVE WASTE; RADIOACTIVITY; RADON; and THREE MILE ISLAND.]

Effects of Radiation Dosage on Human Beings	
Rems	**Effects**
0–25	No detectable effects
25–100	Some blood cells slightly and briefly reduced in number
100–200	Some blood cells suffer longer-term reductions in number; doses greater than 125 rems cause fatigue, nausea, vomiting
200–300	Nausea and vomiting within 24 hours of exposure; latent period (up to 2 weeks) followed by malaise, pallor, diarrhea, sore throat, appetite and weight loss; recovery may take 3 months
300–600	Nausea, vomiting, and diarrhea during first few hours; one-week latent period followed at first by malaise, fever, and appetite loss, then hemorrhage, inflammation of mouth and throat, diarrhea, and weight loss; death may occur within 6 months
600+	Nausea, vomiting, and diarrhea during first few hours; rapid emaciation as early as second week; almost all victims die

Radioactive Fallout

❚❘The debris that rains down from the sky and forms at or beneath Earth's surface following a nuclear explosion. NUCLEAR WEAPONS are extremely powerful. When they explode, they cause shock waves and fires that propel huge clouds of smoke and debris into the air and outward from the site of the explosion. The explosive power of a nuclear weapon comes from the NUCLEAR FISSION of radioactive elements such as URANIUM and PLUTONIUM. A nuclear fission reaction produces a great deal of energy. Thus, radioactive debris from such an explosion is propelled far into the ATMOSPHERE. The dust and debris that fall back to Earth is called radioactive fallout.

Radioactive fallout can also result from accidents at nuclear reactors, such as the accident that occurred at the NUCLEAR POWER plant in Chernobyl, Ukraine, in 1986. Here, the power plant caught fire, sending clouds of radioactive smoke into the air. Since an east wind was blowing at the time, most of the radioactive fallout landed in areas located to the west of Chernobyl. The fallout was detected as far away as in the Scandinavian countries and even in Alaska. [*See also* AIR POLLUTION; CANCER; CHERNOBYL ACCIDENT; NUCLEAR WINTER; RADIATION EXPOSURE; and RADIOACTIVITY.]

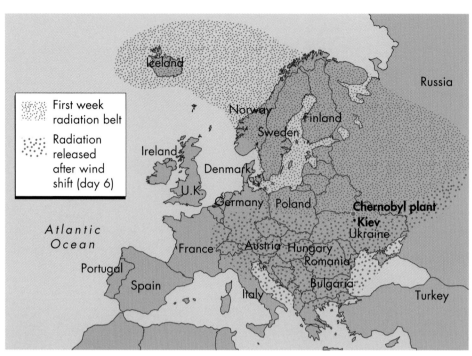

First week radiation belt

Radiation released after wind shift (day 6)

◆ Radiation released into the air by the Chernobyl accident spread to distant areas. The map above shows how the radiation was spread by wind during the first six days.

Radioactive Waste

▮▶ Material produced mainly by nuclear reactors and by MINING activities that emit energy and particles. There are many nuclear reactors in the world. In these reactors, radioactive elements, such as URANIUM, are bombarded by neutrons to bring about NUCLEAR FISSION reaction. During nuclear fission, an atom is split, giving off large amounts of energy as well as some nuclear particles.

Reactors are used to make NUCLEAR WEAPONS, to generate ELECTRICITY, to provide the energy for submarines and ships, and to produce radioactive chemicals for use in research and medicine. These nuclear reactors, as well as various other manufacturing processes, produce waste products that are radioactive. Uranium mines and other types of mines also produce large amounts of radioactive wastes. Radioactive waste is extremely dangerous, and can cause death, CANCER, and birth defects from RADIATION EXPOSURE.

Radioactive waste, like other HAZARDOUS WASTES, must be disposed of in a way that does not threaten the health of humans or other organisms. Radioactive waste is generally classified as low-level waste or high-level waste. High-level wastes emit large amounts of energy and particles for long periods of time. Low-level wastes are not very radioactive or contain radioactive elements that have short half-lives. Many manufacturing processes, research facilities, and medical establishments produce low-level radioactive wastes. Most of the wastes produced by uranium mines are low-level. The most difficult waste to dispose of is the high-level radioactive waste from NUCLEAR POWER plants.

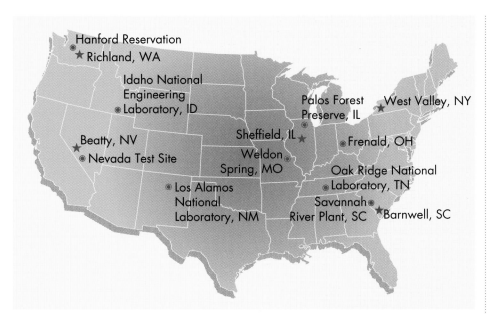

◆ Low-level and high-level wastes from nuclear weapons production and nuclear power plants are stored at sites throughout the United States.

LOW-LEVEL WASTE

Nuclear reactors produce some low-level radioactive wastes. These wastes are hazardous for about 300 years. In the United States, more than 106 million cubic feet (3 million cubic meters) of low-level radioactive wastes (from nuclear power plants, hospitals, and other facilities) have been buried in shallow LANDFILLS. Much of the RADIOAC-TIVITY in these landfills has leaked, and will continue to leak, into the surrounding SOIL and groundwater.

Most nuclear reactors are powered by FUEL rods that contain uranium-235. As little as 2.2 pounds (1 kilogram) of uranium fuel yields as much energy as 2,000 tons (1,800 metric tons) of COAL. Thus, the MIN-ING of uranium damages land less than coal mines because the uranium mines are fewer and smaller. However, uranium must be concentrated before it can be used as a

◆ Burial sites of low-level radioactive wastes produced by commercial activities are shown below.

fuel. To accomplish this, much of the mineral removed from a mine is discarded as mine TAILINGS.

The fuel cycle of uranium produces many environmental hazards. Before the 1970s, uranium tailings were dumped on land. Some were even used as landfill. In Grand Junction, Colorado, tailings were spread over a site before 4,000 houses were built there. Residents were exposed to radiation equivalent to ten chest X RAYS a week. The leukemia rate in Grand Junction was twice that in the rest of Colorado. Over time, the radioactive sand had to be removed at great expense.

HIGH-LEVEL WASTE

Disposing of high-level waste from nuclear reactors has become one of

◆ Yucca Mountain is being considered as a possible place to store radioactive wastes.

the biggest barriers in expanding the use of nuclear power to produce electricity. High-level nuclear wastes remain hazardous for tens of thousands of years. Most people would not want to live anywhere near such wastes. Millions of gallons of high-level liquid wastes are currently stored in steel drums near nuclear reactors. Because these drums corrode, the waste must be repackaged periodically.

YUCCA MOUNTAIN

In 1987, the federal government chose Yucca Mountain, Nevada, at the edge of the Nevada nuclear weapons test site, as the site for the nation's first storage dump for high-level radioactive wastes. The nuclear industry has spent $6.7 billion on studies of the site's suitability for waste disposal. The government plans to seal thousands of tons of used fuel rods and other high-level wastes into steel canisters and store them in a maze of underground tunnels. The tunnels are expected to enclose the radioactivity for 10,000 years.

Although the waste repository is designed to withstand earthquakes up to 7.0 on the Richter Scale, a 1992 earthquake provided supporting evidence for those who oppose using this site for nuclear waste. Evidence shows that earthquakes could damage the canisters and release huge amounts of RADIOACTIVITY into the ENVIRONMENT.

It is likely that a plan to install a high-level waste dump in a less remote area in the United States would be met with even more resistance by the public. [See also CANCER; HAZARDOUS WASTES, STORAGE AND TRANSPORTATION OF; HEALTH AND DISEASE; RADIATION EXPOSURE; and RADON.]

Radioactivity

▌A property of certain elements that give off RADIATION from the nuclei of their atoms. Radiation is emitted as energy and particles.

MINERALS called phosphors glow for some time after they have absorbed sunlight and ultraviolet light, which are forms of energy. In 1896, Henri Becquerel was studying phosphors when he accidentally discovered radioactivity. He found that a sample of the element URANIUM darkened a photographic film even when it had not been exposed to light. From this observation, Becquerel concluded that the uranium was radioactive. We now know that the photographic film was darkened by radiation given off by the uranium. In 1903, Marie Curie discovered two more radioactive elements, polonium and radium.

An atom consists of a nucleus (containing protons and neutrons) surrounded by electrons. The atoms of some elements are very stable. A radioactive element, however, has an unstable nucleus. An unstable nucleus decays or breaks down easily. When a radioactive nucleus decays, it gives off energy and particles.

RADIOACTIVE DECAY

The decay of a radioactive element is commonly described by its half-life—the amount of time it takes half of the sample to change to a stable element. Some radioactive nuclei are more stable than others, so radioactive elements decay at

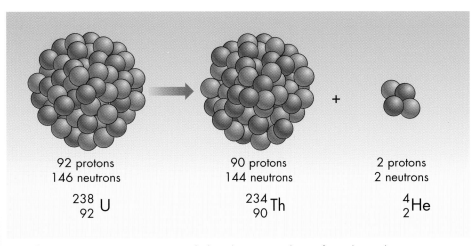

92 protons	90 protons	2 protons
146 neutrons	144 neutrons	2 neutrons
$^{238}_{92}$U	$^{234}_{90}$Th	$^{4}_{2}$He

◆ Radioactive isotopes spontaneously break apart and transform themselves into other elements.

different rates. For example, half a sample of uranium-238 disintegrates in 4.5 billion years. Half a sample of lawrencium-257 decays in eight seconds. The faster an element breaks down, the shorter its half-life. Radioactive decay can be measured using a Geiger counter, an instrument invented in 1908.

ISOTOPES

Most elements exist in more than one form. For example, uranium-238 is one form of the element uranium. Uranium-235 is another form. The two forms are isotopes of the same element. Isotopes are atoms of the same element that have different numbers of neutrons in their nuclei. As a result, the different isotopes of elements have different atomic masses. For example, uranium-238 has an atomic mass of 238 because it contains 92 protons and 146 neutrons. Uranium-235 has an atomic mass of 235 because its nucleus has three fewer neutrons (143 total) than uranium-238.

CARBON DATING

Some elements, including carbon, have both radioactive and nonradioactive isotopes. For example, the

◆ Radioactive isotopes can be detected by an instrument called a Geiger counter.

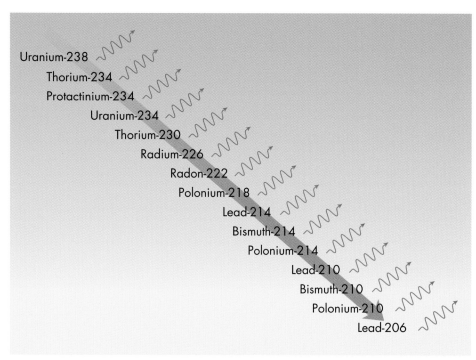

◆ When a uranium isotope decays, it goes through a series of changes and finally stops when it becomes lead.

common and nonradioactive form of carbon is carbon-12. Carbon-12 is very stable. However, carbon-11 and carbon-14 are unstable and are called radioactive isotopes. Since all living things contain carbon in both radioactive and nonradioactive forms, carbon dating can be used to determine the age of biological remains such as a fossil.

Carbon-14 is produced in the ATMOSPHERE and is found in the tissues of PLANTS and animals. As long as an organism is alive, the level of radioactivity in it is constant. However, when an organism dies, carbon-14 decays and changes to a different element. To find out how long an organism has been dead, scientists measure and compare the amounts of carbon-14 and carbon-12 present in an organism's tissues. The half-life of carbon-14 is 5,730

years. By comparing the amount of carbon-14 in the dead organism with the amount constant in the air, the age of the organism can be calculated. Carbon-14 dating is accurate for objects less than 50,000 years old.

RESEARCH AND MEDICAL USES

Ionizing radiation darkens photographic film. Such radiation forms when an atom or molecule takes on an electric charge through the loss or gain of electrons. Ionizing radiation can be detected by a Geiger counter or similar instrument. Thus, radioactive isotopes can be used to determine where a radioactive element is in the body. For instance, the element hydrogen has a radioactive isotope called *tritium*

(H-3). If researchers are interested in finding how rapidly water rises up through a plant stem, they can supply the plant with water containing tritium. Using x-ray photographs or a Geiger counter, the height of the tritium in the stem can be measured at various times. From this data, scientists can calculate how fast the water moves through the stem. [*See also* CHERNOBYL ACCIDENT; HEALTH AND DISEASE; NUCLEAR FISSION; NUCLEAR WASTE; RADIATION EXPOSURE; and THREE MILE ISLAND.]

Radon

▶A colorless, odorless, radioactive gas formed naturally by the radioactive decay of URANIUM and radium contained in rocks. Radon is most concentrated in porous SOILS that lie above rocks that contain uranium. In many regions of the United States, radon gas has been discovered in the basements of homes built on such soils. Radon gets into homes through cracks in foundations and walls. It is sometimes released into the air through water from faucets and showers if the source of water is a well with high levels of radon.

According to the ENVIRONMENTAL PROTECTION AGENCY (EPA), radon causes more CANCER deaths than any other single pollutant except tobacco smoke. The EPA estimates that in the United States, as many as 20,000 lung cancer deaths occur each year as a result of radon. This is a much greater risk than the risk from typical exposures to such environmental health hazards as ASBESTOS, vinyl chloride, PESTICIDES, chloroform in drinking water, or benzene in the air.

Because radon is odorless and tasteless, its presence cannot be detected without the use of special equipment. Short term, inexpensive screening tests are available to measure average radon exposure over time. They provide an indication of the radon levels in a home and tell if further testing is needed. Radon can be removed from air and water with the aid of special venting equipment. [*See also* HEALTH AND DISEASE; POLLUTION; and RADIOACTIVITY.]

◆ Most rain forests are tropical and are located near the equator.

Rain Forest

▌A BIOME that receives an average of 79 inches (200 centimeters) of PRECIPITATION each year in which the dominant PLANTS are trees. Rain forests can be found in temperate areas. The Pacific Northwest of the United States, along with the coast of Alaska, is largely comprised of temperate rain forest. Most of the Earth's rain forests are located within the tropical zone—the region of Earth located between the equator and 30 degrees north and south latitude.

Because of its nearness to the equator, the tropical zone receives the most direct rays of the sun. Thus, areas within the tropical zone are characterized by warm CLIMATES.

The average temperature in these regions is about 68° F (20° C) year-round. In addition, areas within the tropical zone receive from 39 to 177 inches (100 to 450 centimeters) of precipitation each year, with average precipitation around 79 inches (200 centimeters). Because of the warm temperatures, most of this precipitation falls as rain.

RAIN FOREST STRUCTURE

Warm temperatures and abundant rainfall allow the growing season of the rain forest to last most of the year. The dominant form of plant life in the rain forest are its trees, of which there are thousands of SPECIES. The leaves and branches of the tallest trees, which may grow to heights greater than 164 feet (50 meters), form a dense upper **canopy**, or covering, high above

the forest floor, which is called the *understory*. The canopy prevents most of the sunlight that reaches rain forest areas from reaching the forest floor. As a result, the plants that make their home on the forest floor must be adapted to survival in areas where sunlight is extremely limited.

Between the upper canopy and understory levels of the rain forest is a region called the *lower canopy*. This region is made up largely of trees that do not grow to be very tall and those that have not yet reached maturity. The lower canopy receives most of the sunlight that filters through the upper canopy.

RAIN FOREST DIVERSITY

Rain forests cover approximately one-sixth of Earth's land surface.

◆ Each species that lives in the rain forest is adapted to life within a particular section of the forest.

They are considered Earth's most productive land biome. Scientists estimate that roughly 50% of all of Earth's species may inhabit the rain forests.

Plant Life

Among the more common tree species in a rain forest are deciduous trees such as cypress, teak, mahogany, and balsa. Several species of coniferous trees and tree ferns also grow in rain forests.

Because of the limited sunlight reaching the forest floor, vegetation is sparse in the understory. However, some plant species such as FERNS and mosses do thrive here. Other plants that live in the rain forest include various epiphytes—nonparasitic plants that grow on other plants. Epiphytes common to rain forests include a variety of tropical orchids and many mosses. These plants generally grow upon the branches of the tall trees making up the forest canopy.

Animal Life

Like rain forest plants, the animals of the rain forest are adapted to survival in different zones of the forest.

Animals living in the rain forest show great BIODIVERSITY. The diversity of animal life in rain forests is generally attributed to two factors. One is the great diversity of plant life. The second factor is the variety of HABITATS. Habitat variety also results from the diversity of shapes and heights of plants living in the rain forest.

The upper canopy of the rain forest supports a variety of INSECTS, BIRDS, and MAMMALS. These animals all have ADAPTATIONS that allow them to spend their entire lives well above the forest floor. For example, birds and insects have wings that allow them to easily fly from one tree to another in search of food and water that falls upon the leaves of the trees. Mammals that live in this level are adapted to climbing or swinging from one tree branch to another. In addition to movement, these animals also are able to make use of the living space provided by tree branches, leaves, and trunks.

The forest understory also shows great diversity in the species of animals that live there. The high moisture and warmth of the area make the region particularly suitable for **ectothermic** animals such as AMPHIBIANS and REPTILES. In addition, streams and puddles formed by water that falls on the forest floor provide habitat and water resources for the other animal species.

◆ The vegetation of the rain forest in Costa Rica provides a rich and varied habitat for animals.

IMPORTANCE OF RAIN FORESTS

The rain forest is an extremely important biome. First, it is important as a habitat for the vast array of organisms living there. Second, it is important to people because of the many NATURAL RESOURCES they obtain from these areas. These resources include some of the world's most expensive woods, such as mahogany and teak, as well as more common woods, which people living near rain forests often gather for use as FUEL.

The rain forest is also important to people because of its many hidden resources. These resources exist in the many species of organisms people have not yet discovered and identified. It is estimated by some scientists that the number of undiscovered species inhabiting Earth's rain forests may be greater than 50 million. Many of these species may contain chemicals that may someday be made into life-saving medicines. Others may contain materials that can be used in the manufacture of a great many products.

THREATS TO THE RAIN FOREST

At the present time, rain forest regions are disappearing. People have changed and destroyed rain forest habitat either through direct abuse of the resources provided by these regions or by claiming its land for other uses. The removal of large numbers of trees from an area is called DEFORESTATION. People deforest areas for different reasons. Among them is the need to obtain wood for fuel, to create land for use in farming or by GRAZING animals, and to create space to build homes, commercial centers, or other development projects such as roads.

Deforestation is often accomplished through CLEAR-CUTTING or the use of slash-and-burn methods. Both of these practices result in HABITAT LOSS for the organisms that once lived in the cleared area. In addition, both practices may deplete resources needed by plants and animals for their survival, forcing these organisms to move to a new area or die. If enough habitat is destroyed, a species may even face EXTINCTION.

Deforestation has harmful effects on the SOIL in an area, too. Clearing an area of its plant life makes it more vulnerable to processes such as EROSION and DESERTIFICATION. In addition, rain forest soils are not rich in nutrients. Because the region supports so many plants, nutrients returned to the soil through DECOMPOSITION are immediately taken in by the roots of living trees. Thus, the clearing of land in rain forest regions may leave behind a soil that is so depleted of nutrients, it is unable to support the growth of crops.

Some scientists believe deforestation can have harmful effects on global climate patterns. For example, CARBON DIOXIDE is used by trees to carry out PHOTOSYNTHESIS. In this process, OXYGEN that is used by most organisms for RESPIRATION is released into the ATMOSPHERE. When trees are cut down, photosynthesis stops. In turn, the cycling of carbon dioxide and oxygen carried out by these trees also stops. In addition, trees and other plants use carbon dioxide as building material for their tissues. They store carbon dioxide, removing large quantities from the atmosphere. If enough trees are removed from a forest, it may increase the buildup of carbon dioxide in the atmosphere. In addition, the burning down of trees also releases large amounts of carbon dioxide into the air. Many scientists fear that a buildup of too much carbon dioxide in the air will lead to GLOBAL WARMING. [*See also* CARBON CYCLE; CONIFEROUS FOREST; DECIDUOUS FOREST; ECOTOURISM; FOREST; FUEL WOOD; GREENHOUSE EFFECT; GREENHOUSE GAS; LEACHING; OXYGEN CYCLE; SPECIES DIVERSITY; and WILSON, EDWARD OSBORNE.]

◆ White-faced Capuchin monkeys have adapted to living in the canopy of a tropical rain forest.

◆ In many parts of the world, people have turned wild grasslands into rangelands. Instead of supporting wild hoofed animals such as bison or elk, rangelands feed domestic animals such as cattle and sheep.

Rangeland

▎▎Land, mostly GRASSLAND, that is grazed by cattle, sheep, and other domestic animals. Worldwide, people raise cattle, sheep, goats, and other LIVESTOCK on rangelands. This is one way of obtaining food and other products from these animals.

A dry PRAIRIE or a windy hillside that cannot be farmed may be used to support cows or sheep. In fact, the population of such livestock outnumbers the human population in many regions. These animals can survive by GRAZING on local PLANTS. They provide meat, milk, leather, wool, and other products for people. The raising of livestock is very common in much of the world because there are many large areas of grassland that are too dry or too steep for farming. For example, about 70% of the land of the western United States is used as rangeland.

RANGELANDS IN DANGER

Grasses and some other rangeland plants grow back quickly after they are partly eaten. Because of this, a certain number of grazing animals can live and eat on a rangeland year after year, without causing permanent damage to the plant community. Some grazing animals help some plants by spreading seeds from place to place and fertilizing the SOIL with their droppings. This number of animals (whatever it may be for that piece of rangeland) determines the range's CARRYING CAPACITY.

If more sheep, cows, or other livestock are added to the range so that the carrying capacity is exceeded, problems arise. When OVERGRAZING occurs, the rangeland is unable to recover while the animals graze somewhere else. Instead, the tastiest plants constantly lose their leaves, flowers, and stems to hungry animals. After a while, the range begins to look different. Fewer SPECIES of plants that are nutritious and easy to eat remain. Plant species that are tough, toxic, or otherwise hard for grazers to eat become more common. EXOTIC SPECIES may have a better chance to spread on the range. A very badly overgrazed range may lose its plants altogether, with disastrous results.

OVERGRAZING on rangeland is common. Most of the rangeland in the world is overgrazed to some degree. This is true in the United States as well as in other countries. Some rangeland is damaged by people carelessly using land that they do not own. However, people who own their own rangeland often damage it, too.

KEEPING RANGELAND HEALTHY

There are many ways of keeping rangelands healthy and productive. First, of course, the number of grazing animals on the range should not be allowed to be larger than the rangeland's carrying capacity. This is not as easy as it sounds, because the carrying capacity of a range goes up in years of good WEATHER and down in years of little rainfall. The right number of animals in one year can be too many in another year. Also, animals must not be allowed to graze too long in one place. They need to be moved around so that grazed areas can recover. Other ways of caring for rangeland include getting rid of unwanted brush and exotic species of plants, fertilizing the soil, making watering holes, and putting up fences that control the movements of the grazers.

Laws that promote good use of rangeland are also needed. In the United States, the Public Rangelands Improvement Act of 1978 was a step toward better management of public rangelands. However, rangelands in the United States and worldwide are still overused. Improved laws may bring about more repair and protection.

Conserving rangelands takes hard work, expense, and a good knowledge of the rangeland ECOSYSTEM. If efforts are made, rangelands will remain a valuable NATURAL RESOURCE that supports livestock, people, and a number of other species as well. [*See also* AGROECOLOGY; BIODIVERSITY; BIOME; CONSERVATION; DUST BOWL; FIRE ECOLOGY; HERBIVORE; LAND USE; MULTIPLE USE; NATIONAL GRASSLAND; PAMPAS; and PUBLIC LAND.]

◆ Many areas in the western United States are used as rangeland such as this one near Taos, New Mexico.

Reclamation Act of 1902

❚▶A federal act that established a fund to pay for the development of IRRIGATION projects through the sale of PUBLIC LANDS in the western United States. The Reclamation Act of 1902 gave the secretary of the interior broad powers to sell public lands in 16 states of the Great Plains, Rocky Mountains, and Far West. Money from the sale of the lands went into a fund that was to be used to develop irrigation projects that would change dry western lands into agriculturally productive lands. Later, the DEPARTMENT OF THE INTERIOR established the BUREAU OF RECLAMATION, which still controls the construction and operation of a variety of irrigation projects.

Under the Reclamation Act, surveys were to be conducted to identify the most promising areas for irrigation-based agriculture. These lands were also to be sold to people who would live on the land. Buyers were required to develop at least half their property for agriculture and no one could buy more than 160 acres (64.8 hectares). The act also required that the reclamation fund be used for the maintenance and operation of irrigation systems, that most of the fund's money be spent, and that states benefit roughly equally. The act was not intended to affect existing WATER RIGHTS in the West, even though it did.

Irrigation in the West remains a controversial topic, even though agriculture has expanded because of it. Irrigation has brought about various problems. Many irrigation projects were linked to DAMS for HYDROELECTRIC POWER. These projects have affected the long-term health of SALMON populations in many rivers. Irrigation projects that use groundwater have begun to pump more water out of AQUIFERS than can be replenished. Water withdrawals for irrigation restrict the amount of water left for downstream users. In addition, water returning to streams from agricultural areas may be polluted by nutrients and PESTICIDES.

Increases in the populations of agricultural areas in dry western states have placed pressure on water sources needed for drinking water. Irrigation water can also harm dry SOILS by depositing salts from the irrigation water in the upper soil layers. On the one hand, using sprinkler systems for irrigating in very dry CLIMATES can result in the loss of a great deal of water by evaporation. On the other hand, irrigation projects have allowed many western states to become sources of important agricultural products ranging from potatoes and sugar beets to wine grapes. [*See also* AGRICULTURAL POLLUTION; DESERTIFICATION; EROSION; HYDROLOGY; LAND USE; and SALINIZATION.]

◆ Many companies use recycled paper for packaging.

Recycling, Reducing, Reusing

❚▶Methods recommended for the CONSERVATION of NATURAL RESOURCES and the production of fewer waste products. Natural resources are the biotic and ABIOTIC FACTORS of the ENVIRONMENT that are used by living things. FOSSIL FUELS, nutrients, SOIL, WILDLIFE, MINERALS, and water are all examples of natural resources. Some of these resources, such as water and nutrients, are needed by all living things for survival. Other resources, such as fossil fuels and some minerals, are used almost exclusively by people to make their lives more comfortable.

The use of natural resources by people poses two problems for the environment. One problem results from the fact that some natural resources exist in limited supplies or are used by people at a faster rate than the resources can be replaced through natural processes. Such resources, generally classified as NONRENEWABLE RESOURCES, include fossil fuels and minerals.

The second problem associated with the use of natural resources by people is that many of the products made from the resources are later discarded. Products made from organic materials that break down rapidly in the environment do not usually create a serious disposal problem. However, other products, such as PLASTICS and toxic chemicals, are difficult to dispose of in a way that does not harm the environment. To combat both of these problems, environmentalists encourage people to conserve natural resources through the methods of recycling, reducing, and reusing.

RECYCLING

Recycling is a way to reduce the amount of resources used by collecting usable waste materials and reprocessing them to make new products. For example, newspapers and magazines are often collected and processed into new paper products, such as toilet tissue, notebook paper, or new newspapers and magazines.

Recycling benefits the environment in several ways. First, it reduces the amounts of raw materials that must be obtained to make new products. For example, by recycling newspapers and magazines, fewer trees need to be harvested. Second, recycling reduces the amount of waste that must be disposed of in LANDFILLS or other disposal sites. Third, the recycling of materials is often more energy efficient than making new products from raw materials. For example, the recycling of ALUMINUM like that used in the making of beverage

◆ The recycling symbol is used to identify materials that can be recycled. Many consumers look for this symbol when making purchases, to do their part in conserving natural resources.

cans may use as much as 95% less energy than the making of aluminum from its ore. A fourth benefit to recycling is tied to ENERGY EFFICIENCY. By using less energy to make products, fewer pollutants are released into Earth's ATMOSPHERE and water supplies.

Many different materials are currently being recycled. Among the most common are paper, glass, plastics, and motor oil. A variety of metals, including aluminum, COPPER, silver, gold, LEAD, zinc, iron, and steel are also recycled. The recycling of these metals not only requires the use of fewer natural resources in the manufacture of new products but also reduces the amount of MINING that must take place throughout the world.

REDUCING

The concept of reducing is simple: people need to use and dispose of fewer natural resources. The term SOURCE REDUCTION is often used to describe this concept. Source reduction is the lessening of the demand for a resource that results in a reduction in the amount of the resource that is needed to satisfy the demand. The idea of source reduction is often applied specifically to decreasing the amount of waste people produce. Source reduction has all the same benefits to the environment as recycling.

One way to reduce the amount of resources used and the amount of materials people dispose of is to avoid the purchase of materials that are intended to be disposable. For example, using glass or plastic dishes instead of polystyrene foam cups and paper plates reduces the amount of resources needed to

◆ Newspapers are being recycled at the West County Recycling Center in Richmond, California.

◆ After it is sorted by color, glass is crushed, melted, and recycled.

make these products and lessens the amount of GARBAGE created. Similarly, cloth diapers can be used in place of disposable diapers, and mechanical pencils can be used in place of wooden pencils.

Being a careful consumer can also help reduce the unnecessary use of natural resources and reduce garbage production. For example, consumers can try to purchase products that use a minimal amount of packaging material that will be thrown away later. Similarly, buying products that can be used repeatedly or recycled also helps conserve natural resources and reduce waste.

REUSING

The idea of reusing natural resources involves the use of a prod-

◆ The sails on this boat are recycled from plastic soda bottles and plastic car fenders.

uct over and over again either for its original purpose or for a new purpose. Like reducing, reuse may simply involve using reusable rather than disposable materials. However, reuse may also involve using a product in a way other than its original use. For example, reuse may involve using containers that once contained one type of food for the storage of another type of food. For example, if you buy a plastic tub of margarine, you can reuse the tub later as a container for holding leftover food items. Similarly, you can give away clothes you've outgrown to another person who can make use of them. The reuse of items in these ways lessens the demand for natural resources and reduces the amount of waste products that need to be disposed of. [*See also* ALTERNATIVE ENERGY SOURCES; COMPOSTING; CONTAINER DEPOSIT LEGISLATION; CONVENTION ON INTERNATIONAL TRADE IN ENDANGERED SPECIES OF WILD FAUNA AND FLORA (CITES); EARTH DAY; FUELS; FUEL WOOD; WASTE MANAGEMENT; and WASTE REDUCTION.]

◆ When metals are not recycled and left at a dump, valuable resources are wasted.

◆ Flattened car bodies can be recycled instead of being left to rust away.

Renewable Resources

▶Biotic and ABIOTIC FACTORS in the ENVIRONMENT that are formed and recycled through natural processes and are used by living things. A NATURAL RESOURCE is anything in the environment that is useful to living things. Natural resources include matter, such as water, and energy, such as sunlight. All living things must make use of some natural resources to survive.

CLASSIFICATION OF NATURAL RESOURCES

Natural resources are classified into two broad groups: NONRENEWABLE RESOURCES and renewable resources. Nonrenewable resources include all materials and energy sources whose supplies have the potential of being depleted because these materials exist in limited supplies or are used at a rate faster than they can naturally be formed or cycled in the environment. Examples of nonrenewable resources include FOSSIL FUELS and MINERALS. In contrast, renewable resources include all materials and sources of energy that are in abundant supply or are constantly formed or recycled in the environment. Examples of such resources include organisms, such as trees, which recycle themselves through the process of reproduction, and SOLAR ENERGY—the energy in sunlight.

RENEWABLE RESOURCES AND THEIR USE

An ECOSYSTEM provides the organisms living there with all of the things they need for survival. These things include food or nutrients, water, living space, and energy. All of these needs are supplied to organisms in the form of renewable resources. For example, PRODUCERS—organisms that make their own food—obtain their nutrients by combining certain raw materials in the environment either through PHOTO-

◆ Photovoltaic cells capture solar energy and transform it into electrical energy.

SYNTHESIS or chemosynthesis, the production of organic substances in which energy comes from chemical substances. CONSUMERS are organisms that obtain their nutrients by feeding on other organisms. A third group of organisms called DECOMPOSERS obtain their nutrients by feeding upon the remains of once-living organisms. In this process, the decomposers not only obtain the nutrients they need but also return other nutrients to the environment, where they can be used again by producers. All of the materials involved in these feeding relationships—the chemical substances, energy, and organisms—are examples of renewable resources.

CONCERNS ABOUT NATURAL RESOURCES

Although renewable resources are constantly reformed, overuse or misuse of these resources may threaten the survival of organisms. For example, POLLUTION in a variety of forms threatens Earth's air, water, and SOIL resources. Pollution makes air, water, and soil unfit for use by living things. Most of this pollution results from the activities of humans.

Environmentalists encourage people to become aware of the activities that contribute to the pollution problem. People, both individually and in groups, are asked

◆ Geothermal wells can be drilled like oil wells to capture steam and hot water usable for heating and for electricity production.

fossil fuels when these energy resources are used.

To help prevent the overuse or misuse of renewable and nonrenewable resources, most scientists recommend that people become actively involved in conservation practices. These practices encourage recycling, reducing, and reusing, whenever possible, the natural resources that make up the products people use. Conservationists also ask people to reduce the amounts of natural resources they use so that supplies of these resources will be available for use by future generations. [*See also* AGRICULTURAL POLLUTION; AGROECOLOGY; AIR POLLUTION; AIR POLLUTION CONTROL ACT; BIOGEOCHEMICAL CYCLE; BUREAU

◆ Resources are either renewable, nonrenewable, or potentially renewable. Potentially renewable resources can be depleted if we use them faster than natural processes that renew them.

to find ways to carry out these activities in a way that reduces the release of pollutants into the environment. For example, the wood people burn to produce heat and ELECTRICITY and to power machinery releases large amounts of pollutants into the environment. To combat these problems, scientists are seeking ways to make use of ALTERNATIVE ENERGY SOURCES, such as WIND POWER, solar energy, and TIDAL ENERGY. All these energy resources are in abundant supply and are less polluting than fossil fuels. In addition, scientists are working on ways to reduce the amounts of pollutants released into the environment by

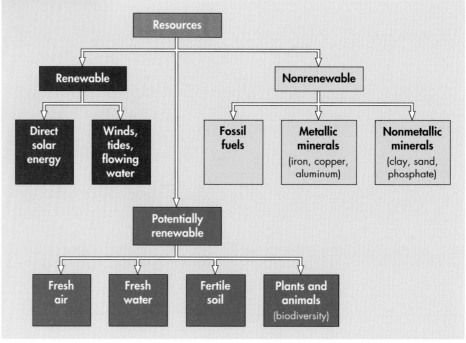

Reptile

❙▶ An air-breathing, ectothermic, scale-covered VERTEBRATE animal, that lives on all continents except ANTARCTICA. Reptiles evolved from AMPHIBIANS fairly late during the Paleozoic Era and quickly came to dominate the lands, seas, and skies of the Mesozoic world. Two major biological ADAPTATIONS— tough skin and water-tight eggs—enabled reptiles to become the first true land animals with backbones. They were also the first vertebrates to fly and the first animals of any kind to walk on two feet.

CLASS REPTILIA

The class Reptilia is divided by zoologists into four living orders.

Order Testudinata (also called Chelonia) includes turtles, tortoises, and terrapins. There are about 260 living SPECIES, with a basic design—a body enclosed in a shell of bony plates covered by horny scales. This order of reptiles predates the dinosaurs. Testudines include CARNIVORES, HERBIVORES, and OMNIVORES. There are land-living, marine, and freshwater species, but all reproduce by laying eggs.

Order Squamata ("scaly") includes two suborders Lacertilia (or Sauria), the 4,100 species of lizard, and Ophida, the 2,800 species of snake. All but 5% of living reptile species belong to this order. The first lizards appeared early during the Mesozoic Era. Lizards usually have legs and always have lidded eyes that can close. Most lay eggs, but some bear live young. Modern lizards range in size from only a few inches in length to the so-called Komodo dragon of Indonesia that can grow up to 10 feet (3 meters) long. Lizards include carnivores, herbivores, and omnivores.

Sometime late during the Mesozoic Era, burrowing lizards gave rise to snakes. Snakes are legless and have lidless eyes that are covered by transparent scales. They range in size from barely 4 inches (10 centimeters) long to the python of Asia and the giant anaconda of South America, both of which may attain lengths of 33 feet (10 meters). Most snakes lay eggs, but some bear live young. All are carnivores. Pit vipers such as rattlesnakes hunt with the aid of a sensory device located between the nostril and eye. The special nerves detect infrared RADIATION; that is, the nerves are sensitive to heat, including the body heat of other animals.

The order Crocodilia includes approximately 24 carnivorous species of crocodiles, alligators, caimans, and gavials (gharials). All

◆ The sidewinder rattlesnake has a rattle on the end of its tail and a poisonous bite delivered through a pair of fangs. The snake sometimes strikes without giving a warning rattle.

◆ One-month-old sea turtles are shown above. In the middle is the Komodo Dragon and at the bottom is an alligator swimming toward her newly hatched offspring.

crocodilians reproduce by laying eggs in sand or mounds of rotting vegetation. One parent, though not always the mother, generally remains close by to guard the nest and unearth the hatchlings when they emerge from the eggs. Adults of some species transport their newly hatched young (by mouth) to special nursery areas in swamps.

The order Rhynchocephalia ("beak head") consists of one living species, the tuatara (*Sphenodon punctatus*). It is all that survives of a once-large group dating from the early Mesozoic Era. The tuatara exists only on a few small islands in the Bay of Plenty, New Zealand, where it has protected status. It grows to 2 feet (60 centimeters) in length and feeds on INVERTEBRATES and small vertebrates. A third eye, located on top of the tuatara's head, detects infrared light.

REPTILES' PLACE IN THE WORLD

Reptiles contribute to the structure and functioning of ECOSYSTEMS throughout the world's temperate and tropical regions. They are often important members of FOOD CHAINS. Small lizards, for example, are PREDATORS of INSECTS and other small invertebrates and serve in turn as food for BIRDS and MAMMALS. Snakes are important predators of rats, mice, and other rodents. A single snake may eat nearly 150 mice in a six-month period.

Reptiles sometimes live in commensal or protocooperative relationships with other organisms. In a commensal relationship, interaction between species benefits one species without affecting the other. Certain land tortoises, such as gopher tortoises, create burrows that other animals use. On its remote islands, the tuatara commonly shares its burrow with a bird, the petrel. In a protocooperative relationship, both species benefit, though neither is dependent on it. African crocodiles tolerate the attentions of the crocodile bird, which picks insect pests from crocodile hides and may even feed on leeches around crocodile mouths.

THREATS TO REPTILE POPULATIONS

Throughout the world, destruction of HABITAT, MARINE POLLUTION, and overharvesting for food, leathers, medicines, and the exotic-PET TRADE threaten to extinguish reptile species. Crocodilian populations have recovered from a worldwide decline only in those regions where leather-trade harvesting is regulated.

In some places, snakes are killed out of fear and dislike or for "sport." Annual events called rattlesnake roundups have become popular attractions in some Texas communities. Before each roundup, rattlesnakes are flushed from their burrows and collected so that they may be hunted by tourists. The meat is served up as fast-food, the hides fashioned into belts, boots, and the like, and the skulls and rattles become trophies and trinkets. Competing communities have vied for the state legislature's official recognition as rattlesnake-roundup capital of Texas.

◆ The Gila monster is a poisonous lizard living in deserts in the United States and Mexico.

Texas adopted the rare horned lizard as its Official State Reptile in 1993. However, PESTICIDE use has reduced the horned-lizard habitat in the American Southwest.

All populations of all species of SEA TURTLES are in decline. Sea turtles are protected by law in some parts of the world, but the protection does not extend as far as the animals travel. The turtles' yearly returns to the same nesting beaches generally take them to countries where they are freely hunted. Many other turtles die from ingesting oil-tars and PLASTICS in GARBAGE dumped at sea. Still others drown when they become trapped in fishing nets. In some waters, nets equipped with devices called *turtle excluders* are required by law. Continued public education has helped protect the remaining few sea turtles. [*See also* CLEAN WATER ACT; COMMENSALISM; DARWIN, CHARLES ROBERT; ENDANGERED SPECIES; FISHING, COMMERCIAL; GALÁPAGOS ISLANDS; MARINE PROTECTION, RESEARCH, AND SANCTUARIES ACT; MUTUALISM; NATURAL SELECTION; OCEAN DUMPING; and SYMBIOSIS.]

Reservoir

▶ A basin, either natural or artificial, where a large quantity of water is collected and stored. Lakes are natural reservoirs. Artificial reservoirs are made by building DAMS across valleys.

Reservoirs supply communities with water for drinking, IRRIGATION, manufacturing, and the generation of HYDROELECTRIC POWER. Reservoirs also control **drainage** in an area. Millions of people throughout the world depend on dams for their survival.

DRAWBACKS TO RESERVOIRS

While artificial reservoirs solve problems, they also create problems. When a river is dammed, rising waters drown the dry land behind the dam and destroy the existing ECOSYSTEM. For example, Africa's fertile Volta Valley now lies beneath that continent's largest artificial lake. Ecosystems downstream from dams are disturbed by declines in water flow and by changes in water temperature and water quality.

Damming a river traps not only water but also the silt the water carries. Ordinarily, silt is deposited as SEDIMENTS in vast **deltas** at the mouths of rivers. However, when a river's current is slowed by a dam, silt settles out of the water and collects behind the dam, gradually filling the reservoir basin. When Boulder Dam on the Colorado River

THE LANGUAGE OF THE ENVIRONMENT

deltas more or less triangular tracts of land created from sediments deposited at the mouths of rivers.

drainage passage of water over and through the land surface.

♦ The reservoir behind Grand Canyon Dam is called Lake Powell.

was completed in 1936, the capacity of its reservoir, Lake Mead, was 32,471,000 acre-feet, or about 10 trillion gallons (38 trillion liters), of water. By 1970, silt had reduced Lake Mead's capacity by almost 2 million acre-feet. All reservoirs silt up, but the process is most rapid in deforested regions where much TOPSOIL is washed into rivers. The Dominican Republic's 80,000-kilowatt Tavera Hydroelectric Project was completed in 1973. By 1984, silting had reduced its reservoir's capacity by 40%.

POLLUTION OF RESERVOIRS

Like other bodies of water, reservoirs are vulnerable to POLLUTION. Reservoirs are fed by rivers and streams into which contaminants may have been released. As water from a reservoir flows into and out of a community, chemicals are added during the purification process. WASTEWATER containing DETERGENTS goes out to SEPTIC TANKS, where it leaches into the ground and eventually finds its way to rivers and streams that feed reservoirs downstream.

WATER RIGHTS

Both in the United States and worldwide, as the demand for fresh water overtakes and exceeds the supply, rivers will continue to be dammed. Disagreements over WATER RIGHTS are likely to erupt into open conflict when dams are built on rivers that supply water to more than one country. About 40% of people rely on water that originates in another country. The demand for

◆ The Hoover Dam is on the Arizona-Nevada border and controls floods of the Colorado River. It supplies water and electric power.

water can only increase as the human population increases.

Fights over water rights can occur within countries, too. In the United States, Arizona and California have disagreed over water rights to the Colorado River for many years. The Colorado is one of the largest bodies of water in the mostly dry western United States, flowing for 1,450 miles (2,333 kilometers), but so much water is diverted along the way that the river becomes a tiny stream or trickle as it winds through the deserts of Mexico, toward the Gulf of California. [See also AGRICULTURAL POLLUTION; AQUIFER; BONNEVILLE POWER ADMINISTRATION; CHLORINATION; DEFORESTATION; DISSOLVED OXYGEN; FISH LADDER; FLOODPLAIN; HABITAT LOSS; HYDROLOGY; LEACHING; SAFE DRINKING WATER ACT; SALINIZATION; SEDIMENTATION; TENNESSEE VALLEY AUTHORITY;

THERMAL WATER POLLUTION; WATER, DRINKING; and WATER POLLUTION.]

Resource Competition
See COMPETITION

Resource Conservation and Recovery Act (RCRA)

▮ Legislation that establishes guidelines for how SOLID WASTES are to be transported, stored, and disposed of. The Resource Conservation and

Recovery Act (RCRA) was enacted in 1976 to address a growing problem in the United States—how to safely dispose of the millions of tons of solid and HAZARDOUS WASTES that were being generated nationwide each year. RCRA is administered by the ENVIRONMENTAL PROTECTION AGENCY (EPA).

Hazardous wastes include a variety of chemicals such as PESTICIDES, heavy metals, polychlorinated biphenyls (PCBS), dyes, cleansers, and RADIOACTIVE WASTES, to name a few. Hazardous wastes can be extremely harmful to people and the ENVIRONMENT. They can directly poison animals and PLANTS, causing disruptions in ECOSYSTEMS. When such wastes enter the drinking water supply, they can also affect human health. The RCRA contains strict requirements for how industries must handle and dispose of their hazardous waste.

REQUIREMENTS OF THE RCRA

As public awareness about the environment grew in the 1970s, it became clear to Congress that some action had to be taken to address the problem of hazardous waste disposal. The RCRA has three main goals:
1. to protect humans and WILDLIFE from the harmful effects of hazardous wastes;
2. to reduce the amount of solid waste generated in the United States and to conserve NATURAL RESOURCES; and
3. to reduce or eliminate the production of hazardous wastes.

The RCRA has four distinct programs for achieving these goals.

◆ The Resource Conservation and Recovery Act of 1976 established guidelines for the handling and disposal of solid wastes.

already polluting the environment. In 1980, Congress passed the COMPREHENSIVE ENVIRONMENTAL RESPONSE, COMPENSATION, AND LIABILITY ACT (CERCLA), often called the SUPERFUND Act. Under this act, a large trust fund, or "superfund" was set aside to clean up areas contaminated by hazardous wastes. These laws also give the EPA the legal right to sue the owners of hazardous waste sites and make them pay for cleanup procedures.

The RCRA has been amended several times, most recently in 1984. These amendments, known as the Hazardous and Solid Waste Amendments (HSWA), expanded the scope of the RCRA and made handling and disposal requirements even stricter. [*See also* CARCINOGEN; HAZARDOUS MATERIALS TRANSPORTATION ACT; HAZARDOUS SUBSTANCES ACT; HAZARDOUS WASTE MANAGEMENT; HEAVY METALS POISONING; LOVE CANAL; TOXIC SUBSTANCES CONTROL ACT (1976); TOXIC WASTE; and WATER, DRINKING.]

Under the first program, individual states are to develop plans for managing and disposing of nonhazardous wastes, such as paper, cardboard, glass, ALUMINUM, PLASTIC, and other GARBAGE generated by homes and businesses.

The second program is the main body of law in the RCRA. It addresses the problem of hazardous wastes. Under these laws, producers of hazardous wastes must keep complete records of how their wastes are handled from the time they are made to the time they are disposed of in a regulated hazardous waste disposal facility.

The third program in the RCRA relates to the problem of underground storage. This program contains laws requiring industries to install new, leakproof tanks to prevent dangerous chemicals from leaking into groundwater supplies. If the wastes and other chemicals stored in the tanks ever cause a problem, the producer is legally responsible for cleaning them up and fixing the problem.

The fourth program in the RCRA is also the newest. It concerns the problem of MEDICAL WASTE and how it should be disposed of.

RCRA AND THE SUPERFUND ACT

The RCRA established rules and regulations for how hazardous wastes should be handled. However, it did not address the problem of hazardous wastes that were

Respiration

▶ The chemical process whereby molecules of food are broken down inside an organism's body for use by the organism. Respiration in this sense, known as *cellular respiration*, takes place at the microscopic level within all living cells. At the macroscopic level, respiration refers to the act of breathing, or the way many AEROBIC organisms obtain OXYGEN and release CARBON DIOXIDE. In both cases, respiration involves the exchange of gases between an organism and its ENVIRONMENT.

CELLULAR RESPIRATION

During cellular respiration, an organism releases the energy stored in food. For most PLANTS and animals, this takes place through a complex series of steps during which organic molecules, such as the sugar glucose, are chemically combined with oxygen. The resulting energy is used by the organism to carry out its life functions. Carbon dioxide and water are given off as waste products.

In green plants, respiration is the reverse of PHOTOSYNTHESIS. However, respiration occurs in both the light and the dark. During the day, photosynthesis takes place more quickly than respiration, so plants give off more oxygen than they take in. At night, plants only take in oxygen and give off carbon dioxide.

BREATHING

Most people equate respiration with breathing—the process whereby many animals take in oxygen and give off carbon dioxide and water vapor. Each time you breathe in, or inhale, oxygen from the air enters your lungs. It then passes into the bloodstream where it is available for use in cellular respiration. Each time you breathe out, or exhale, the waste products of cellular respiration are returned to the environment. Because breathing and cellular respiration go hand in hand, smoking, AIR POLLUTION, and anything that affects the flow of oxygen into an organism can affect

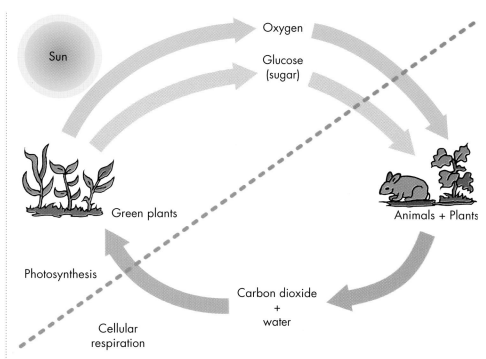

◆ During photosynthesis, green plants use energy from the sun to convert carbon dioxide and water into glucose. During cellular respiration, organisms combine glucose with oxygen, releasing energy and giving off carbon dioxide and water.

this vital gas exchange. [*See also* CARBON CYCLE; DEFORESTATION; HEALTH AND DISEASE; and OXYGEN CYCLE.]

Restoration Biology

▌Applied science that attempts to restore damaged ECOSYSTEMS to their predisturbed states. Restoration biology projects can involve turning an old farm into a PRAIRIE or converting low-lying fields into WETLANDS.

Restoration biology began early in the twentieth century. About this time, scientists began to recognize the environmental problems resulting from industrialization and agriculture. Aldo LEOPOLD was one of the leading biologists in this field at the time. He studied environmental problems by performing small-scale experiments in his gardens. Leopold argued that damaged ecosystems could be restored to functioning ecosystems through direct human intervention.

RESTORATION BIOLOGY PRACTICES

Restoration of an ecosystem requires the contributions of scientists working in many fields. These scientists include climatologists, geologists, engineers, biologists, and chemists, among others. Usually, the first step

◆ Restoration biologists attempt to restore damaged ecosystems to their predisturbed states.

in any ecological restoration project is to change damaged SOILS into soils suited to plant growth. After plowing and furrowing to increase the water retention of soil, fertilizers are often laid down to replenish nutrients in the soil. Once the soils are ready for the introduction of PLANTS, scientists seed the area with key plant SPECIES.

Restoration depends upon the ecological process of SUCCESSION. As key species grow, they create soil conditions suitable for the growth of other plant species. Some key species grow better next to particular plants, so additional plants are sometimes planted as necessary.

Scientists recognize that ecosystems cannot be restored to their exact, predisturbed states. However, by reintroducing many of the native animal and plant species, a stable, self-perpetuating BIOLOGICAL COMMUNITY may be achieved. Animal colonization of a restored ecosystem is usually left to nature. However, suitable animal species can be introduced when necessary.

The successful restoration of an ecosystem can take many years. Restoration sites are closely monitored. Over a period of months or years, unwanted species are removed from a restoration site. The actions of private citizens can also be important in a restoration project. Many communities help provide the funding for long-term projects or volunteer in cleanup and restoration efforts.

A RESTORATION BIOLOGY SUCCESS STORY

The science of restoration biology is still in its infancy. The outcome of a restoration project is almost always uncertain. However, most result in a vast improvement to the ENVIRONMENT. For instance, one of the greatest environmental success stories in the United States involved restoration efforts at Lake Washington near Seattle. In the 1950s, scientists discovered that SEWAGE emptied into the lake had caused severe damage to the ecosystem. Over the years, government agencies, environmental planners, and public citizens studied the problems and developed efforts to correct them. Part of the plan involved the building of a new sewer system in 1963 that emptied treated wastes into the OCEAN rather than the lake. Today, Lake Washington is cleaner than it has been since scientists began studying the lake in the 1930s. [*See also* ABIOTIC FACTORS; BIODIVERSITY; BIOGEOCHEMICAL CYCLE; BIOREMEDIATION; BUREAU OF RECLAMATION; CLIMAX COMMUNITY; CONSERVATION; ECOLOGY; ENVIRONMENTAL EDUCATION; GREAT LAKES; LAND STEWARDSHIP; OFFICE OF SURFACE MINING, RECLAMATION, AND ENFORCEMENT; RESOURCE CONSERVATION AND RECOVERY ACT (RCRA); and WILDLIFE REHABILITATION.]

Restoration Ecology

See RESTORATION BIOLOGY

Rio Conference

See UNITED NATIONS EARTH SUMMIT

Riparian Land

▎Land at the edge of a river or stream, including land that may be covered with water during floods. Riparian land, especially that in semiarid regions, provides HABITAT to hundreds of PLANT and animal SPECIES.

Plant species often include a variety of grasses and trees that are

◆ Riparian land provides a habitat for hundreds of plant and animal species, such as the elk and the gray wolf.

referred to as riparian trees. Animal species may include INSECTS, especially those whose larvae require water for development, as well as migratory BIRDS. In addition to providing habitat, riparian land serves an important role in preventing soil EROSION. The roots of plants living in this area hold SOIL in place, which then helps to prevent soil erosion during periods of flooding. [*See also* RIPARIAN RIGHTS.]

Riparian Rights

▌The rights of owners of lands on the banks of rivers and streams. Under American and international law, people who own land on the edges of rivers have certain rights to the river water and to the SOIL and MINERALS on the bottom of the river. They have the primary right to withdraw water for reasonable use. Generally speaking these rights are:

1. Use of the water near RIPARIAN LANDS for bathing and domestic use. Sometimes this also includes the right to remove water from the river for IRRIGATION.

2. The right to build a dock from the bank out into the river far enough to keep a boat.

This current property law in the United States was developed as part of English law dating back to colonial times. It was designed to ensure landowners access to waterways, such as rivers, in an agrarian or farming society. Landowners could use the water as long as they did not interfere with the quality or quantity of the water flow. [*See also* RIPARIAN LAND.]

Risk Analysis

▌Assessment of the chances of an undesirable outcome from a particular course of action. Risk analysis involves adding the likelihood of exposure to what is known about the risk of being exposed.

Before deciding how best to solve environmental problems, governments and individuals take into account the cost of a project, the benefits to be gained from it, and the risks of not undertaking the project. The most important risk usually consided is the risk of death of humans or other organisms.

Economists point out that it is often worth spending large sums to prevent environmental damage. For example, suppose a government agency is planning to work to control a particular kind of AIR POLLUTION. In estimating how much to spend controlling air pollution, a government should add the value of plants and WILDLIFE killed, the cost of purifying fresh water polluted from the air, the cost of cleaning and renovating buildings damaged by air pollution, the cost to human health, and all the other indirect costs of air pollution.

As we discover more and more indirect costs of POLLUTION, we find that pollution costs people huge sums. The United States spends almost $100 billion a year on pollution control. Economists are convinced that spending much more would be well worth the money.

When people decide how much to spend controlling air pollution, they perform a COST-BENEFIT ANALYSIS. Such an analysis balances the cost of the action against the benefits we expect from it. The cost-benefit analysis of solving an environmental problem tends to depend on who is doing the analysis. To an industry, the cost of pollution control or a safety measure may seem all-important. Industry officials often argue that the cost will be passed on to the consumer or taxpayer, who will refuse to pay it.

In analyzing risk, two factors must be considered together. The first is the risk from a particular event, such as smoking a pack of cigarettes or having an accident at a NUCLEAR POWER plant. The second is how likely individuals are to be exposed to the risk. Even if the risk to human health from smoking or from a nuclear accident is extremely large, someone who never smokes or is not exposed to second-hand smoke and lives 1,000 miles (1,650 kilometers) from the nearest nuclear reactor may be in no danger from either source. [*See also* ENVIRONMENTAL EDUCATION; ENVIRONMENTAL ETHICS; ENVIRONMENTAL JUSTICE; RADIATION EXPOSURE; and RISK ASSESSMENT.]

Risk Assessment

▌The process of deciding how great a chance a particular action has of producing an unfortunate result. To make a decision regarding how much to spend on environmental problems such as AIR POLLUTION, a COST-BENEFIT ANALYSIS is performed. This is done to balance the cost of the action against the benefits people expect from it.

Risk of an undesirable outcome is one of the costs of any action. Risk assessment helps people create cost-effective laws to protect human health and the ENVIRONMENT. It is important that the risk actually

Rankings of Some Environmental and Health Risks by Scientists and the General Public		
	Scientists	**General Public**
High Risk	1. Global warming	1. Chemical waste disposal
	2. Indoor air pollution	2. Water pollution
	3. Exposure to chemicals in consumer products (pesticides, food additives, plasticizers, etc.)	3. Chemical plant accidents
	4. Surface water pollution (risk to the environment)	
	5. Pesticides	
Low Risk	1. Hazardous waste sites	1. Indoor air pollution
	2. Underground storage tanks (mainly of petroleum)	2. Exposure to chemicals in consumer products
		3. Global warming
		4. Pesticides

posed by a hazard equals the risk the public perceives. If the public perceives the risk as being greater than it really is, legislators may enact laws that cost more than necessary to achieve the goal.

PERCEPTIONS OF RISK AND GAIN

How good is the general public at risk assessment? Studies suggest that people are erratic in their judgments. They are often over-influenced by media coverage of risks. The way a risk is presented affects how people assess it. Losses seem more important than gains, so people are more willing to gamble to avoid losses than to achieve gains. In addition, few people understand the rules of probability (chance), which is what determines risk. Because people do not understand probability, certainty is very appealing. In one study, people playing a simple game of chance would pay more money to increase their chance of success from 90% to 100% than they would to increase it from 60% to 70%, though the 10% increase in performance gives the same result in both cases.

The desire of people for absolute certainty tends to make low probabilities seem higher and high probabilities seem lower than they actually are. For example, a high probability of success, such as 85%, may seem insufficient. A low probability of failure, such as 5%, may seem unacceptable.

The most important risk people have to consider is the risk of death. When asked about causes of death, however, most people overestimate the death rate from sensational causes such as homicide, tornadoes, botulism, and drowning. The inaccurate data

Relative Risks to Human Welfare	
Relatively high-risk problems	Habitat alteration and destruction Species extinction and loss of biodiversity Stratospheric ozone depletion Global climate change
Relatively medium-risk problems	Herbicides/pesticides Toxins and pollutants in surface waters Acid deposition Airborne toxins
Relatively low-risk problems	Oil spills Groundwater pollution Thermal pollution

occurs because people get most of their information from the media, which tends to emphasize the spectacular.

INADEQUATE DATA

People are more prepared to take a risk when they think they understand or can control the situation than when they do not. For instance, many people smoke cigarettes but are frightened of flying.

In one study, respondents said that the risks of NUCLEAR POWER were involuntary, uncontrollable, unknown, inequitably distributed, likely to be fatal, potentially catastrophic, and dreaded. AUTOMOBILES evoked few of these concerns, though they pose many of the same risks. The driver often has no control over a road accident. There is little justice in the distribution of road deaths and crashes kill many people. Part of the judgment in this case involves the riskiness of assessing the risk.

People consider technologies such as nuclear power and GENETIC ENGINEERING to be highly risky, partly because there is no way to assess their risks accurately. People do not trust the judgment of the "experts" who tell them what they think the risks may be. This grows out of people's yearning for certainty—they prefer knowing that coal-generated ELECTRICITY on average kills a certain number of people each year to not knowing how many people might be killed by a nuclear accident. [*See also* ENVIRONMENTAL EDUCATION; ENVIRONMENTAL ETHICS; ENVIRONMENTAL IMPACT STATEMENT; ENVIRONMENTAL JUSTICE; ENVIRONMENTAL POLICY; and RISK ANALYSIS.]

River Basin

The area of land from which water flows into a network of streams and rivers that, in turn, flow to the OCEAN as one large river. People often think of a river as a line. However, a river is actually more like the veins of a leaf, with dozens of little streams coming together to form the river. All the waterways that run into the river are its **tributaries**. For instance, the Wabash River is a tributary of the Mississippi

THE LANGUAGE OF THE ENVIRONMENT

permeable allowing water or other liquids to easily pass through.

tributaries streams that feed a larger stream or lake.

◆ Smaller streams joining larger streams and rivers make this extensive rootlike pattern in western Colorado.

◆ The Green River flows into the Colorado River as it flows south into the Gulf of California.

River, which the Wabash joins at Cairo, Missouri.

A river basin is also called the *drainage basin,* or WATERSHED, of the river. The river basin of the Mississippi is huge. It occupies about one-third of the United States east of the Rocky Mountains.

A river's water comes from PRE-CIPITATION—rain, snow, or sleet. In regions where the SOIL is not **permeable,** and in urban areas, rain that falls on most of the river basin may run into the river. In humid regions, most of the water that runs into a stream comes from rain that falls on low areas near the stream. Much of the precipitation that falls on land never reaches a river because it is absorbed by PLANTS. DEFORESTATION (the cutting down of a large number of trees in an area) almost always increases the surface RUNOFF.

People usually think of runoff from land, but runoff may also come from underground. An underground AQUIFER may drain into a river by way of a spring that brings water to the surface. An aquifer may also drain directly into the sea. About one-third of the discharge of the world's rivers comes from underground water, which is often the only thing that keeps a river running between periods of rain and snow melt. The remaining two-thirds comes from surface runoff.

The edges of the river basin are the highest points on the hills and mountains surrounding the river and its tributaries. If you stand on top of a ridge on the edge of a river basin, you may be able to see streams on either side of the ridge that flow in opposite directions. This happens because they are part of different river basins and empty into the oceans in different places. For example, there are many places along the Appalachian Mountains where you can see, on one side of a ridge, a stream that is a tributary of the Mississippi, which empties into the Gulf of Mexico. On the other side, you can see a stream that empties into the Atlantic Ocean. [*See also* FLOODPLAIN.]

Rodenticide

▶A poison used to kill rats, mice, gophers, moles, and other animals classified as rodents. Rodents are often a problem for humans, particularly rats, which cause millions of dollars worth of damage to buildings and food supplies annually. Each year, mice, gophers, and moles cause damage to lawns and gardens. Rodents carry dangerous diseases such as **tularemia**, **rabies**, food poisoning, and **rat bite fever**. Each year, rats bite about 150,000 people. Thus, eliminating rats from places inhabited by people is an important part of disease control. Poisoning the rats and rat-proofing buildings is the most successful method of eradication.

KINDS OF RODENTICIDES

There are three basic kinds of rodenticides—*acute*, *chronic*, and *fumigants*. Acute and chronic rodenticides are put in foods as baits.

Fumigants are sprayed into the air. Acute rodenticides include zinc phosphate and cholecalciferol. These chemicals kill by damaging the animal's vital organs such as its kidneys, liver, and heart. Strychnine is an acute rodenticide that damages the animal's nervous system.

Acute poisons cause "bait shyness" in rats. At first, a rat will eat only a small amount of a new food. If the rat becomes sick within the next day, it will not eat more of the food. However, bait shyness is not a problem when chronic rodenticides are used. These slow-acting poisons do not cause illness for several days. Thus, the animals may continue to feed on the bait. Chronic rodenticides act by destroying the clotting ability of the blood. This causes the animal to die of internal hemorrhage.

Fumigant poisons are sprayed into contained areas such as buildings. These poisons, used to get rid of large numbers of rodents, include methyl bromide and cyanide. Since these poisons can kill people, only professional fumigators who understand the hazards associated with use can apply these poisons.

◆ Rats destroy many tons of grain each year.

PROBLEMS WITH RODENTICIDES

Care must be taken when using rodenticides. Most are poisonous to people, domestic animals, such as pets and farm animals, FISH, and BIRDS. For this reason, many people are beginning to focus their methods of PEST CONTROL on the INTEGRATED PEST MANAGEMENT (IPM) concept. Integrated pest management makes use of methods other than poisoning whenever possible. These alternative methods include recruiting natural enemies to remove the unwanted pests and lacing bait with substances that retard growth and reproduction. Thus far, integrated pest management methods have been unsuccessful in eliminating rodents, such as rats. The recruiting of a family cat is also ineffective. The practice also exposes the cat to the diseases carried by the rodents. [*See also* ADAPTATION; BIOLOGICAL CONTROL; FEDERAL INSECTICIDE, FUNGICIDE, AND RODENTICIDE ACT (FIFRA); HEALTH AND DISEASE; and PESTICIDE.]

Roosevelt, Franklin Delano (1882–1945)

❚❘Thirty-second president of the United States, whose long and eventful administration (1933–1945) spanned the nation's worst depression and greatest war. Assisted by Secretary of the Interior Harold Ickes, Roosevelt oversaw the establishment of numerous federal con-

servation programs, including the Civilian Conservation Corps, the

TENNESSEE VALLEY AUTHORITY, and the SOIL CONSERVATION service. During his term, the building of water conservation and HYDROELECTRIC POWER plants was authorized, including the Central Valley Project in California, and the Grand Coulee Dam in Washington. In addition, the Jackson Hole National Monument and Olympic and King Canyon national parks were established by the NATIONAL PARK SERVICE.

Roosevelt, Theodore (1858–1919)

❚❘The twenty-sixth president of the United States, whose "Square

Deal" program included a CONSERVATION movement. Theodore "Teddy" Roosevelt was born into a wealthy New York City family in 1858.

As a young child, Roosevelt developed a strong interest in science and nature. During this period, he spent much of his time collecting natural objects, such as rocks, bones, and small animals. He kept these objects at his home for scientific study. The young Teddy made careful observations and measurements of the objects in his collection and kept careful records of his observations in his diary. His cousins dubbed the collection, which was kept in Teddy's dresser, as the "Roosevelt Museum of Natural History."

Theodore Roosevelt was a sickly and frail child. However, to overcome these problems, Roosevelt began to exercise vigorously. By the time he entered Harvard

University in 1876, Roosevelt had built himself up enough to participate on the college boxing team.

LIFE IN POLITICS

After graduating from Harvard, Roosevelt obtained his law degree from Columbia University. Soon after, he began his political career in 1881 in the New York State legislature. Following his term in office and the death of his first wife, Roosevelt held a variety of other positions both in and out of politics. Then, in 1900, Roosevelt was elected to the position of vice president of the United States under William McKinley.

Only one week into his term as president, William McKinley was assassinated and Roosevelt was sworn in as President of the United States. Four years later, he ran for and was elected to the presidency of the United States based largely on a program that he called the "Square Deal."

One aspect of Roosevelt's Square Deal program was an emphasis toward improving the ENVIRONMENT. This emphasis dealt mostly with the conservation of NATURAL RESOURCES, particularly the nation's FORESTS and WILDLIFE.

To achieve this goal, Roosevelt set aside 150 million acres (60.7 million hectares) of timber land in the West and 85 million acres (34.4 million hectares) in Alaska as national forest land. The land was placed under the control of the U.S. FOREST SERVICE.

At about the same time, the RECLAMATION ACT of 1902 was passed. This act authorized use of money obtained through the sale of government land in 16 states to be used for the development of IRRIGATION projects. As a result of this law, many DAMS were built to help provide the water needed to people living in semiarid regions.

In 1908, Roosevelt expanded on his conservation efforts through the creation of the National Conservation Commission. The purpose of the commission was to develop policy to preserve the nation's resources for the future. At the same time the national commission was created, governors in 41 states created state conservation commissions to focus on the same issues.

POST-PRESIDENTIAL YEARS

After serving two terms as president, Roosevelt declined to run for a third term. He spent much of his time traveling the world and HUNTING in remote areas. Following a year-long trip to Africa, Roosevelt published a book in 1910 called *African Game Trails*. Three years later, Theodore Roosevelt, still an avid naturalist, made an expedition into the jungles of Brazil. During this expedition, Roosevelt injured his leg on a rock while exploring the River of Doubt. The wound quickly became infected and the condition was made worse with Roosevelt's development of malaria. For the next few years, these two conditions caused Roosevelt's health to deteriorate. He died in his sleep at his home in Oyster Bay, New York, in 1919. [*See also* FORESTS.]

Runoff

◗ Water from rain, snow, or melting ice that is not absorbed into the ground but flows across Earth's surface into rivers and streams. PRECIPITATION that falls on land may trickle down into the SOIL and rocks to become part of an underground AQUIFER, or it may become runoff, which flows across Earth's surface into a river, stream, or lake.

Heavy runoff can cause EROSION and POLLUTION. After a heavy rain, and especially in an area where much of the vegetation has been removed, runoff carries with it large amounts of soil, causing erosion. The runoff also carries fertilizer and PESTICIDES from farms and yards as well as oil, pet feces, litter, and dust from city streets. These all mix into the runoff, which ends up in waterways, contributing to the WATER POLLUTION problem.

DEFORESTATION has been shown to cause a large increase in surface runoff. One study in North Carolina showed that the amount of water running off the surface to a stream increased more than 1,000% when all the FOREST in the stream's WATERSHED was cut down. The runoff returned to normal within two years due to regrowth. However, the critical problem was that most of the TOPSOIL was lost before the new plants became established. [*See also* AGRICULTURAL POLLUTION; AGROECOLOGY; AGROFORESTRY; CONTOUR FARMING; DAMS; DREDGING; SEDIMENT; and SEDIMENTATION.]

◆ Runoff along this dirt road has caused deep gullies in the road and has washed soil into a nearby river.

S

Safe Drinking Water Act

◗ Law passed by Congress in 1974 and amended in 1987 to protect the public from contaminated drinking water supplies. The Safe Drinking Water Act gave responsibilities for clean water suppplies to the U.S. ENVIRONMENTAL PROTECTION AGENCY (EPA) and to the states.

Under the act, the EPA establishes standards for maximum contaminant levels (MCLs) in drinking water supplies. The MCLs are guidelines for allowable levels of specified toxic chemicals. The EPA requires that municipal water supplies be monitored regularly for these chemicals. If any toxic chemical is found to exceed its allowable level, the water supply is closed until adequate water purification procedures or other alternatives are adopted.

The EPA also protects underground sources of drinking water through the monitoring of wells that are used as storage sites of HAZARDOUS WASTE. In the process called **deep-well injection**, many liquid wastes are being pumped deep into the ground. Some of these wastes contain high amounts of toxic chemicals. **Groundwater** above such wells must be monitored to make sure no toxic waste has escaped its well. If it has, the groundwater could carry the chem-

◆ This water purification system is adjacent to the multipurpose reservoir in Duncan, Oklahoma.

icals to new locations, possibly polluting drinking water supplies.

Responsibility for enforcing drinking water standards, monitoring water supplies, and reporting requirements rests with the states, which must also establish guidelines for deep-well injection of wastes. The EPA awards grants to be used for supporting state public water supply, **wellhead** protection, and underground injection programs.

BOTTLED WATER

Safe drinking water is not only desirable, but absolutely necessary.

Yet, there has sometimes been sharp disagreement about the act's effectiveness.

Some people believe it does not go far enough. These people may argue that the act should be expanded to contain provisions for monitoring bottled water. Many people think bottled water is pure. Many now drink such water because they are concerned about levels of contaminants in municipal water supplies. However, in 1991, many samples of bottled water were shown to contain higher levels of contamination than did municipal water supplies.

PRIVATE WELLS

The Safe Drinking Water Act does not contain any provision for the monitoring of toxic chemicals in private wells. Thus, contamination of a private well may not be recognized unless the water has a strange taste, a bad smell, or until health problems occur that are attributable to the use of the water.

A county health department will test water in a private well at the property owner's request. However, if a contaminant detected in the water does not exceed its specified MCL (the level of contamination found to be "safe"), no action is taken by the county health department. The property owner is left to decide what must be done. If dangerous levels of contamination are found, the well may be closed. If this happens, there is no payment to the property owner, who then must get water from another source or else pay for purification measures.

Some people question the need for a Safe Drinking Water Act. Many feel it is an intrusion by federal government into local affairs. Others argue that the act should be repealed because communities in the United States have never found contamination at levels that require remedial action. Thus, those communities are being forced by law to spend money to protect themselves against a nonexistent hazard. Monitoring is especially costly for small towns of only a few hundred households, which are required to do the same monitoring as large cities. [*See also* CLEAN WATER ACT; WATER POLLUTION; WATER QUALITY STANDARDS; and WATER TREATMENT.]

◆ The EPA requires cities to monitor their water regularly to detemine whether or not it contains more than the allowable concentration of toxic chemicals.

Salinity

The degree of SALINIZATION, or accumulation of salts, in SOIL or water. Salts occur naturally in all soils, but plants cannot grow in a locality where salinity is too high. Sometimes the salts become so concentrated that the ground forms a hard crust. IRRIGATION increases salinity because it uses river water or GROUNDWATER, which are saltier than rainwater; thus, more salt is left behind when the water evaporates. Continued irrigation can raise salinity in a locality. Dissolved salts carried upward by rising water accumulate in the soil as the water evaporates. Soil can become alkaline if such deposits contain many carbonates. [*See also* SALINIZATION.]

Salinization

The accumulation of salts in SOIL. SALINITY is a measure of the amount of dissolved salts in a substance. All water, both fresh and marine, contains some dissolved salts. The difference between marine, or salt, water and fresh water lies in the amounts of dissolved salts the water contains. For example, OCEAN water has a salinity of about 3.5%. This water contains approximately 35 grams of salt in every liter of water. In contrast, water that is considered fresh has a salinity of about .05%. This water contains only about 0.5 grams of salt in every liter of water.

Sensitive	Tolerant	Moderately Tolerant	Highly Tolerant
Field bean	Barley (grain)	Soybean	Wheat (grain)
White clover	Sugar beet	Rice	Cotton
Dutch clover	Barley (hay)	Oats (hay)	Alfalfa
Carrot	Rye (hay)	Corn (forage)	Wheat (hay)
Radish	Asparagus	Lettuce	Tomato
Grapefruit	Date palm	Corn (sweet)	Beet
Orange		Potato	Kale
Lemon		Pecan	Pomegranate
		Peach	Fig
		Apricot	Olive

◆ The table shows soil salt tolerance of various crop plants. Many of the plants under the heading "Sensitive" would have difficulty growing in soil with a high salt content.

The types of salts most often present in water include the sodium chloride that people use as table salt, as well as calcium sulfate, magnesium sulfate, and calcium carbonate. These salts enter the water supply when they dissolve in rainwater that washes over and runs off the land.

The salts in fresh water do not normally create a problem for people who drink the water or for organisms in the ENVIRONMENT. However, if too many salts accumulate in soil, it becomes unsuitable for the growth of PLANTS. This process is called *salinization.*

Salinization is most common in areas where farmers must rely heavily on IRRIGATION to grow crops. In such areas, water containing dissolved salts is often pumped from lakes or streams onto farm fields. If the fields lack good drainage, water collects in the upper layers of the soil. Much of the water not taken in by plants evaporates from the soil. However, the salts contained in the water are left behind. Over time, these salts build up in the upper soil layers.

If too much salt accumulates in the soil, the soil cannot support the growth of plants. Without plants, TOPSOIL becomes vulnerable to EROSION. Over time, the loss of topsoil combined with the dryness and barrenness of the soil may result in DESERTIFICATION. Thus, an area that was once covered by crops may become a DESERT. Extensive soil damage and desertification resulting from salinization has occurred in Pakistan, Iraq, Egypt, Argentina, and in the southwestern United States. Many other areas are at risk of losing productive cropland because of salinization. [*See also* LEACHING; SALINITY; and SALTWATER INTRUSION.]

Salmon

▶ A family of **carnivorous** FISH, most of which hatch in cold freshwater streams, spend their adult lives in the OCEAN, and then return to their hatching sites to breed. Salmon are important to both the commercial and recreational fishing industries. However, salmon are threatened by HABITAT LOSS and overfishing.

Different SPECIES of salmon exist in North America, Europe, and Asia. The North American species include the Atlantic, coho, pink, chinook, chum, and sockeye salmon, and some species of trout.

Many of these species are *anadromous.* Anadromous species live most of their lives in the ocean, but breed in freshwater streams. Newly hatched salmon remain in a stream for a short time before migrating downstream to the ocean.

There, they may spend many years feeding and growing to adult size. When mature and ready to breed, a salmon swims back into the stream where it hatched. It can recognize its own stream by using its keen sense of smell. A salmon may swim hundreds of miles to return to its **spawning** grounds. Some species make the trip only once and die as soon as they have spawned. Other species may return home more than once. This may seem to be a hard life, but it can give salmon the best of both worlds. They can grow large in the ocean where food is plentiful but breed in a far-off stream, where there are fewer PREDATORS to eat their young.

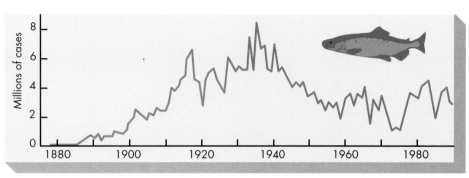

◆ Since about 1975, the United States and Canada have been trying to restore the salmon population. This graph shows the rise and decline of the salmon harvest in Alaska. One case of salmon holds 48 pounds (22 kilograms). Note that these figures are for the salmon that were caught. The number of salmon actually in the population is harder to estimate.

◆ Salmon may swim hundreds of miles to return to the stream where they spawn. Some species die soon after; others live to return several times.

SALMON FISHING

For centuries, salmon have been highly valued by people. They are popular with sport fishers. They are nourishing and tasty. They can also be caught in large numbers when they enter streams and rivers to spawn. There once were so many of this fish that overfishing them did not seem possible.

However, salmon populations were significantly reduced by human activities. Overfishing has reduced their numbers so much that commercial salmon fishing has almost stopped in some areas. In some spots the fish are already gone. This is true of Atlantic salmon in some eastern rivers of North America. In other areas, fishing has been greatly restricted to allow the stocks to recover. This is the case of the Pacific salmon industry of the Alaskan coast.

People have also changed the rivers and streams where salmon breed. Many rivers are now blocked by DAMS. Some streams are now warmer and more full of SEDIMENTS because FORESTS near their banks have been cut. Warm, sediment-filled spawning grounds are unhealthy for salmon. Pollutants also damage spawning areas.

THE FUTURE OF SALMON

Even though many people are concerned about salmon, we still do not understand all the reasons for changes in their populations. Scientists and fishers continue to study salmon. Meanwhile, many efforts are underway to preserve or increase salmon populations. Laws limiting commercial fishing are becoming more strict. Hundreds of thousands of young salmon are raised in the shelter of fish **hatcheries** before being released into the wild. FISH LADDERS are installed to help salmon get over the dams. Many rivers are now less polluted than they were in the recent past, so that salmon may be returned to them. In addition, many of the salmon we eat are grown in captivity instead of being taken from the wild. [*See also* AQUACULTURE; FISHING, COMMERCIAL; and FISHING, RECREATIONAL.]

Salt Marsh

◗ An area of low, flat, poorly drained land that is flooded daily or occasionally by salt or brackish waters. Salt marshes are examples of coastal WETLANDS. Like all marshes, they are covered with thick mats of grasses and small grasslike PLANTS. Salt marshes are common along the eastern coast of the United States, inside barrier beaches, within ESTUARIES, and at the mouths of rivers.

DIVERSITY AND IMPORTANCE OF SALT MARSHES

Salt marshes are among the most productive and diverse ECOSYSTEMS. The constant flow of water from the OCEAN, river, or other body of water into the salt marsh provides high concentrations of nutrients for organisms to thrive on. Salt marshes

◆ Salt marshes are coastal wetlands located near oceans, bays, lagoons, rivers, and other sources of salt or brackish water.

are generally teeming with life. Grasses, cattails, and small shrubs such as sedges and rushes are common plant SPECIES. Typical animals in salt marshes include small INVERTEBRATES such as worms; INSECTS; and crustaceans, such as the fiddler crabs that burrow into the thick mud where they feed upon the decaying organic matter. These species, in turn, attract many small MAMMALS and BIRDS, which feed upon the smaller animals during rest stops or breeding periods.

In addition to providing a home and sanctuary for countless plants and animals, salt marshes and other coastal wetlands are important to the ENVIRONMENT. Most notably, salt marshes and other wetlands help filter pollutants from the oceans, rivers, and streams that feed them. Salt marshes also act as important barriers by absorbing large amounts of water that might otherwise move inland and cause flooding.

THREATS TO SALT MARSHES

Salt marshes are threatened by many human activities. WATER POLLUTION caused by RUNOFF of PESTICIDES, oil, household chemicals and other pollutants is one threat to salt marshes. The main threat to salt marshes and other coastal areas is more direct and has resulted from the way people viewed these ecosystems. Until recently, salt marshes and other wetlands were thought of as little more than worthless breeding grounds for insects such as mosquitoes. Because of this view, many wetlands, particularly the coastal salt marshes, have been drained, filled in, and built upon.

◆ Some salt marshes have rocky bottoms where green algae and scallops live.

◆ Fiddler crabs are common to many salt marshes.

In fact, since the 1600s, about 50% of all wetlands have been destroyed to make room for farms, roads, houses, and other buildings.

Today, the remaining salt marshes of the United States are protected under the CLEAN WATER ACT. The part of the Clean Water Act that protects wetlands has been criticized by supporters of wetland development. Such criticism led to political battles that forced Congress to redefine wetlands in recent years. This change has caused a loss of thousands of acres of coastal wetlands. [*See also* BARRIER ISLANDS; EVERGLADES NATIONAL PARK; HABITAT LOSS; MANGROVES; NONPOINT SOURCE; OCEAN DUMPING; SALINITY; and WETLANDS PROTECTION ACT.]

Saltwater Intrusion

❙❙The movement of salt water into freshwater AQUIFERS. The movement of salt water into an aquifer can make the water unusable by organisms. Such intrusion is most often a problem in coastal regions.

Much of the water people use for drinking, cleaning, industry, and farming is groundwater—water located beneath Earth's surface. Much of this groundwater is stored in aquifers. Though an aquifer can contain a great deal of water, it does not usually form an underground lake. It is more like a soggy sponge. Certain layers of the ground—for example, layers of sandstone—become soaked, or saturated, with water. This water trickles into the ground from rain that falls on it or from another underground location. If a well is driven into an aquifer, the groundwater can be pumped to the surface and used by people.

FRESH WATER NEXT TO SALT WATER

In coastal areas, aquifers may contain salt water as well as fresh water. Most of the time the salt water and fresh water separate because fresh water is less dense than salt water. As a result, fresh water forms a layer on top of the salt water in the aquifer. Thus, wells in coastal areas (or even on islands in mid-ocean) can give fresh water, as long as the wells do not go too deep or remove too much water.

WHY SALT WATER INTRUDES

A large portion of the human population lives in coastal areas, and the coasts are getting more and more crowded all the time. To meet the demands of large populations, more groundwater must be pumped from coastal aquifers. Shortages of rain or snow can also shrink the supply of fresh water available in aquifers.

As fresh water is removed, but not replaced from an aquifer, sea water easily oozes into the aquifer to take its place. Sometimes, this saltwater intrusion reaches the level of the wells that provide water to a community. When this happens, the well water is contaminated and becomes too salty for people to use.

Saltwater intrusion also affects coastal wilderness areas. One such area is the EVERGLADES NATIONAL PARK. In this area, the supply of fresh groundwater has been reduced because of human demands for water outside the park boundaries. This has caused salt water to intrude into lakes, ponds, and rivers that are used as HABITAT or drinking pools by WILDLIFE.

REDUCING SALTWATER INTRUSION

One way to deal with saltwater intrusion is to get water from areas such as inland RESERVOIRS. However, transporting water long distances can be expensive and is not always possible. Another way to prevent saltwater intrusion is to reduce the

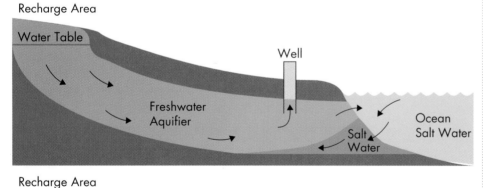

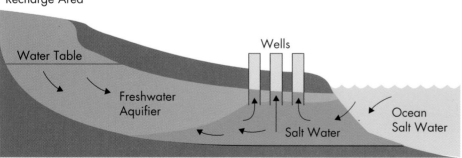

◆ Salt water creeps underground when too much fresh water is removed from the ground in coastal areas. As a result, wells may pump up water that is too salty for drinking, for farming, or for using in industry.

amount of fresh water being used in or near coastal communities.

There are many methods for conserving water in households, on farms, and in industry. If these methods are used, a freshwater aquifer may be able to refill at a rate that is about equal to that of water removal. Such a refill rate helps prevent salt water from entering an aquifer. [*See also* SALINITY; WATER, DRINKING; WATER QUALITY STANDARDS; WATER RIGHTS; WATER TABLE; and WATER USE.]

Savanna

▶ A GRASSLAND characterized by wet and dry seasons and the presence of some scattered trees as well as grasses. Savannas are located in many tropical regions of the world. These BIOMES form in regions where the average temperature is between about 63° and 83° F (15° and 30° C) and the average rainfall is less than 20 inches (50 centimeters) a year.

Usually the rainfall in a savanna is seasonal—it falls only part of the year. In a savanna, a warm, wet season is generally followed by a cool, dry season, and then a hot, dry season. These CLIMATE conditions cause the SOIL of many savannas to be fine and sandy.

LIFE IN A SAVANNA

In a savanna, water for PLANTS is in short supply. The plants that are best able to survive in savannas are grasses. The roots of grasses

◆ The African savanna has a larger number of species of grazing mammals than anywhere else in the world. The zebra and water buffalo are two examples.

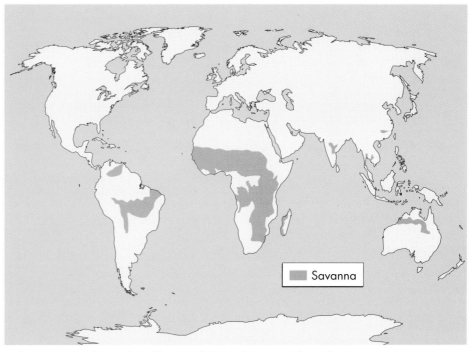

◆ Savannas are located in Africa and South America, where they support large populations of grazing mammals.

form thick networks that are well adapted to getting water from the soil. The stems and leaves of a grass may die in the dry season. However, its roots stay alive and send up new leafblades during the next rainy season. Because grasses can grow back from their roots again and again, they are well adapted to surviving disturbances that occur in savanna ECOSYSTEMS. One of these is fire. Fire often sweeps across savannas, burning the grasses to the ground. The other condition often survived by savanna grasses is the great number of GRAZING animals that live on savannas. In the savannas of Africa, these grazers include large animals such as zebras, antelopes, and gazelles, and millions of smaller animals such as rodents and INSECTS.

Many savannas have enough rainfall to support the growth of trees as well as grasses. In these savannas, there are scattered trees. Though scattered trees sometimes grow in the grassy region of a savanna, they are often killed by fires and droughts. In some places, they are also browsed upon by large animals such as giraffes and ELEPHANTS. All of these events help prevent the wetter savannas from being taken over by FORESTS.

Some of the best-known savanna animals are the large HERBIVORES of the African grasslands. These animals include the giraffe, buffalo, many kinds of zebra, antelope, and gazelles. Many of these animals migrate long distances each year, seeking areas where there is rainfall and new grass. Smaller animals that cannot travel may survive the dry seasons by becoming **dormant** in a process called **estivation**.

MANAGING SAVANNAS

Savannas can be very productive ENVIRONMENTS. The lush growth of grasses during the wet season can feed a large number of animals. Because of this, some savannas are used for grazing cattle to provide meat and other products for people.

Like other grassland HABITATS, savannas must be protected from OVERGRAZING. Overgrazing can happen if too many cattle are kept on a savanna or if a wildlife preserve is too small for the grazing animals to move from one area to another during the year. Overgrazing can upset the balance between the grasses and the woody plants such as shrubs and trees. If grass is eaten

◆ An elephant takes advantage of the shade provided by an acacia tree in the Masai-Mara National Park in Kenya.

away, it may be replaced by other kinds of plants. When this occurs, a savanna may develop into some other type of biome such as a woodland or a thorn forest. The new habitat may produce less grass for cattle or other large herbivores to eat. Thus, careful management of savannas is important to all SPECIES that use them, including cows, people, and WILDLIFE. [*See also* ADAPTATION; FIRE ECOLOGY; MIGRATION; NATURAL DISASTERS; PRAIRIE; RANGELAND; and TROPICS.]

Scrubber

▶A device for controlling AIR POLLUTION that helps remove environmentally damaging substances from smokestack emissions. When power plants and factories burn FOSSIL FUELS, such as oil and COAL, many pollutants are produced. These chemicals include HYDROCARBONS, SULFUR DIOXIDE, and NITROGEN OXIDES. Scrubbers help to reduce the amounts of these substances released into the ATMOSPHERE by "scrubbing" smoke to make it cleaner.

Scrubbers work in two ways. In coal-burning power plants, scrubbers are used to wash PARTICULATES (solids) from smoke. This is done by spraying water at very high speeds into incinerator emissions. In this way, the scrubbers force particulate matter from the smoke.

Scrubbers also absorb harmful gases from incinerator emissions. As smoke flows through the scrub-ber, the gases dissolve in a liquid solvent, much like sugar is dissolved in a cup of hot tea. The mixing of the incinerator smoke and solvent washes polluting chemicals out of the smoke. Air pollutants commonly controlled in this way include sulfur dioxide, nitrogen oxides, and hydrocarbons.

Scrubbers are effective at reducing air pollution. The CLEAN AIR ACT requires many industries to use scrubbers and similar pollution-control devices. While such devices do not completely remove pollutants, they do reduce the amounts of pollutants released into the air. [*See also* ACID RAIN; CATALYTIC CONVERTERS; GREENHOUSE GAS; OZONE; PRIMARY POLLUTION; and SMOG.]

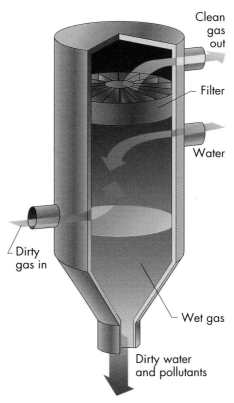

Clean gas out

Filter

Water

Dirty gas in

Wet gas

Dirty water and pollutants

◆ Scrubbers work by removing polluting chemicals from smoke.

Seabed Disposal

▶The practice of getting rid of high-level RADIOACTIVE WASTE or TOXIC WASTE by sending it to the bottom of the OCEAN. Finding storage places for all the dangerous wastes people produce is an endless task. Certain wastes, such as high-level radioactive waste and highly toxic waste, should be kept well away from homes, food supplies, and supplies of fresh water. Some of these wastes must be stored safely for as long as ten thousand years.

There are few sites on land in the United States where waste can be stored with no chance of contaminating the human ENVIRONMENT someday. Even the safest sites that have been suggested are rejected by many of the people who would be living near them.

Some people have looked to the ocean floor as a possible site for waste disposal. Nuclear wastes, for example, could be encased in glass and stainless steel and deposited beneath SEDIMENTS on the ocean floor. Waste sites must be carefully chosen to avoid parts of the sea where there is plate movement and where there are many marine organisms. Wastes could be dropped or buried in parts of the deep ocean in which few organisms live and which might remain undisturbed for thousands of years. Backers of this method of disposal believe that even if waste leaked from containers, it might be diluted by enough ocean water so as not to pose a threat to living things.

◆ Pollution from this canal is clearly visible where it empties into a large body of water.

WHY NOT USE THE SEABED?

Many people oppose seabed disposal. One argument states we do not know enough about the oceans to predict what will happen to waste on its bottom. Will it stay where we put it? Will it leak into the environment? Will it begin to build up in the bodies of marine animals that we eat? Our radioactive wastes may not add much to the natural level of RADIATION in the whole ocean. But would large amounts of waste in one spot cause local damage? Not everyone agrees with the answers to these questions.

Another argument against seabed disposal is that the seabed does not "belong" to anyone in the same way that land does. Many countries share ocean resources, so it is hard to enforce general rules about using the ocean. As a result, individuals, companies, or governments may be tempted to dump wastes in an irresponsible way.

The United States already has plans to dispose of highly dangerous wastes on land instead of in the ocean. However, the safety of sea disposal versus land disposal is still a question in the minds of many scientists and environmentalists. [*See also* HAZARDOUS WASTES, STORAGE AND TRANSPORTATION OF; MARINE POLLUTION; OCEAN CURRENTS; and OCEAN DUMPING.]

◆ Old tires are disposed of on the seabed in the hope that fish might use them for protection.

Seals and Sea Lions

▶M̲arine MAMMALS with paddlelike feet for swimming. Along with walruses, seals and sea lions belong to a family of CARNIVORES called *pinnipeds*. There are about 33 SPECIES of seals and sea lions in the world. They come in many sizes. The harbor seal is about the size and weight of a large dog. A male elephant seal, at about 2,000 to 4,000 pounds (900 to 1,800 kilograms), may be heavier than several horses.

Seals and sea lions live in many places. Some live in warm regions like the GALÁPAGOS ISLANDS and Hawaii. Most species, however, are found in the Arctic Ocean or in ANTARCTICA. They feed on a variety of marine animals. Some species feed on FISH or squid; some eat crustaceans such as shrimp and crabs; some catch and feed on penguins or other seals.

Seals and sea lions look somewhat different from each other. In seal species, the ear is a small hole on the side of the head. Sea lions have a small flap of skin around the ear hole. The flap makes their ears more visible. Another distinguishing trait of seals is that they can only drag their hind flippers underneath their bodies while sea lions use the flippers as feet.

Both seals and sea lions are well suited for spending most of their time in the water. Like other aquatic mammals, they must breathe air to keep from drowning. However, they can stay underwater for long periods and at great depths. The elephant seal is able to

◆ Sea lions sometimes come ashore to rest on piers such as these at Pier 39 in San Francisco.

dive 5,000 feet (1,500 meters) below the surface and go without breathing for up to two hours. Seals and sea lions can even fall asleep at sea, without breathing water by accident. Some of these mammals live far from land for weeks or months at a time.

Seals and sea lions must come ashore to mate and have young. Depending on the species, they may use beaches, rocky islands, or ice floes as breeding sites, or *rookeries*. Rookeries are usually crowded with bulls, cows, and newborn pups during the breeding season.

PEOPLE, SEALS, AND SEA LIONS

Seals and sea lions have been valuable to people for their oil and their fine fur coats. HUNTING them for profit began hundreds of years ago. Since then, species after species of seals and sea lions have been greatly reduced. For example, about two million Alaskan fur seals were killed in the Pribilof Islands during the late 1700s and early 1800s. About one million fur seals were killed in or near the Antarctic during the same period. Starting in 1911, the killing of seals and sea lions began to be controlled by international agreements to limit hunting. Some species then increased again, but some are still threatened or ENDANGERED SPECIES.

Today, changes in HABITAT and MARINE POLLUTION are more dangerous to many pinniped species than is hunting. People compete with seals and sea lions for space along

◆ Sea lions hunt for their food. These sea lions are shown pursuing a school of fish.

shorelines. People also harvest more and more of the seafood that seals and sea lions need to eat. In addition, seals and sea lions are entangled, injured, or killed by items such as nylon fishing nets and plastic rubbish.

In 1988, a disease known as distemper killed many seals in the North Sea, the Baltic Sea, and the North Atlantic Ocean. It was suspected that high levels of POLLUTION had made the seals weak and more likely to become ill. Since then, several countries have agreed to reduce the amount of pollution sent into the North and Baltic Seas. [*See also* ANIMAL RIGHTS and MARINE MAMMAL PROTECTION ACT.]

Sea Turtle

▶A marine REPTILE that belongs to the order Chelonia. Compared to other types of turtles, sea turtles are quite large, have less bulky shells, and have feet shaped like paddles.

There are seven SPECIES of sea turtle: the green turtle, the flatback, the leatherback, the loggerhead, the hawksbill, the olive ridley, and the Kemp's ridley. The hawksbill, one of the smaller species of sea turtle, may weigh less than 100 pounds (45 kilograms) when full grown. The largest species, the leatherback, may reach a weight of 1,200 pounds (540 kilograms).

Most sea turtles live in tropical and subtropical OCEANS. Some, such as the green turtle, feed on marine grasses and other PLANTS. Others, such as the loggerhead, are PREDATORS of animals such as crabs, clams, FISH, and jellyfish.

All sea turtle species lay their eggs on land. Female sea turtles may swim hundreds of miles to bury their eggs on sandy beaches. During its life, a sea turtle faces the dangers of two worlds. It begins life on a beach, where land predators such as foxes and gulls feed on eggs and young turtles. Then it moves into the ocean and must face other predators such as sharks and marine MAMMALS. Sea turtles have been surviving these dangers for millions of years. However, they also face other dangers that may cause them to become extinct.

SEA TURTLES AND PEOPLE

People have hunted sea turtles and taken their eggs for hundreds of years. The eggs are a valuable food. The turtles themselves are hunted for meat, shells, and leather. Egg-collecting and hunting have driven many populations of sea turtles to a point where they are nearing EXTINCTION.

People also destroy the nesting HABITAT of sea turtles. Many beaches have been taken over by people for recreation or building. Other beaches are overrun with predatory animals that live near people, such as dogs, cats, gulls, and raccoons. Newly hatched baby turtles on beaches use sunlight reflecting off the water as a beacon that directs them to the sea. The hatchlings are sometimes confused by lights from highways and cities. Instead of going straight into the sea, some of these young turtles crawl in the wrong direction and die.

In the sea, sea turtles are often killed when they become tangled in the nets of commercial fishing boats. Still others are killed by trash and pollutants that are dumped into the ocean. Plastic bags are especially dangerous. Turtles mistake them for jellyfish, swallow them, and die when the bags block their digestive systems.

PROTECTION FOR SEA TURTLES

In many parts of the world, sea turtles are considered ENDANGERED SPECIES. One of the most important ways of protecting them is to protect the beaches where they nest. Worldwide, many organizations and governments are involved in this effort. One example of a protected beach is the Tortuguero National Park in Costa Rica, which is the world's largest green turtle refuge.

◆ Sea turtles are considered an endangered species in many parts of the world.

Other efforts protect adult turtles. If shrimp fishing boats put turtle-excluding devices on the open ends of their shrimp nets, the turtles have a way to escape, and fewer sea turtles are caught and killed in the nets. Laws and international agreements, such as the CONVENTION ON INTERNATIONAL TRADE IN ENDANGERED SPECIES OF WILD FAUNA AND FLORA (CITES), help to keep turtle products, such as health and beauty lotions, from being sold between countries. However, at this time, sea turtles are still in danger. Much more needs to be done to protect them in many parts of the world. [*See also* FISHING, COMMERCIAL; MARINE POLLUTION; OCEAN DUMPING; and WATER POLLUTION.]

Sediment

▌Solid material temporarily suspended in water or air that has fallen onto an underlying surface. Material is constantly eroded, or carried away, from every surface. For example, dead cells are brushed off your skin, hair falls from your head and from the bodies of animals. In addition, wind carries away TOPSOIL from unplanted fields and water carries away soil and particles of rocks. All these materials, and many more, are carried away by air and water. Eventually, these materials separate from the substance carrying them and fall onto a surface as sediment.

Sediment from the air may fall in your house as dust on a table. Sediment in water may be deposited at the bottom of a river. [*See also* BIOGEOCHEMICAL CYCLE; EROSION; PARTICULATES; SEDIMENTATION; and WEATHERING.]

Sedimentation

▌The act or process of depositing solid pieces of organic or inorganic material. Air, water, and ice carry particles of material with them when they move. When your desk is dusty, it is because some of the particles carried in air have settled out of the air, forming a layer of dust. Another example of sedimentation is seen when SOIL from a field is blown away and falls as SEDIMENT on a nearby road.

Why does sedimentation occur in some places more than in others? The answer is related to movement. When air or water moves, it can carry more sediment than when it is still. Thus, a strong wind can blow large amounts of sand into the air, to create a sandstorm. However, when the wind dies down, the sand falls to the ground. Similarly, water tumbling down a waterfall may carry a large amount of material with it. The material will not sediment out of the water until the water slows further downstream.

Water is more dense than air, so it can carry larger particles. Particles of organic matter, soil, volcanic debris, and rock are carried along by moving water. Some of the particles will eventually drop to the stream or ocean bottom as sediment. This may form sandbars that can change the course of a river or alter water depth in the ocean.

Moving ice has even more power than moving water. GLACIERS may carry huge rocks along with them. The rocks sediment out when the ice stops moving or melts. Most of the rocks in the soil and lakes in the northern part of the United States are actually sediment brought from farther north during the last ICE AGE.

DAMS sometimes "silt up." This occurs when a lot of sediment is deposited in the RESERVOIR immediately behind the dam. The sediment is dropped at the place where the water stops moving. Sometimes the sediment grows so heavy it puts great pressure on the dam. If heavy rainfall adds water to the reservoir, adding even more pressure on the dam, the dam may burst. In the United States, this happened to a dam at Buffalo Creek in West Virginia in 1972. The reservoir behind the dam contained water full of sediment from coal MINING. After three days of rain, the dam collapsed completely, causing a flood that killed 118 people and resulted in $65 million in damages.

Rivers carry some of their sediment from land to the ocean. More sediment forms in the ocean itself as organisms die and their teeth, shells, and other hard parts fall to the bottom. Sediment covers most of the ocean floor. In some places it is more than 3,300 feet (1,000 meters) thick. In most places, it averages about 1,650 feet (500 meters).

Over long periods of time, sediment may be changed into sedi-

◆ The community below a dam at Buffalo Creek suffered considerable damage when the dam silted up and burst in a flash flood.

mentary rocks, such as limestone and sandstone. For instance, limestone is rock formed on the ocean floor by the pressure of water and sediment on the calcium carbonate shells of dead marine organisms.

Geological processes, such as volcanic eruptions, movement of the continents, and earthquakes, may heave rocks up from the bottom of the sea until they are on dry land. Because of this process, sediments found under the ocean are much younger than comparable deposits on land. The oldest marine sediments discovered so far are from the Jurassic period (about 150 million years ago). The oldest rocks on land are about 3.8 billion years old. [*See also* AIR POLLUTION; BIODEGRADABLE; COMPOSTING; CONTINENTAL DRIFT; CORAL REEF; DECOMPOSITION;

DREDGING; EFFLUENT; EROSION; FLOODPLAIN; PARTICULATES; POLLUTION; RIVER BASIN; RUNOFF; SALINIZATION; SLUDGE; SOIL CONSERVATION; TAILINGS; VOLCANISM; WEATHER; and WEATHERING.]

Septic Tank

▶An underground storage container for solid organic waste and liquid matter that is decomposed by microorganisms. People who live in rural areas often discharge their SEWAGE into septic tanks. In a septic tank, anaerobic DECOMPOSERS break down organic matter that is contained in sewage.

Inorganic solids that are not broken down by decomposers sink to the bottom of a septic tank as SLUDGE. After organic matter is decomposed and inorganic solids settle as sludge, the water that remains flows through an outlet pipe to a area called a *leach field*. This water that drains into a leach field is rich in mineral nutrients and is gradually dispersed into the SOIL where it is absorbed by PLANTS. Water released into the leach field may still contain some organic wastes. These wastes may be decomposed by organisms in the ENVIRONMENT. The sludge that

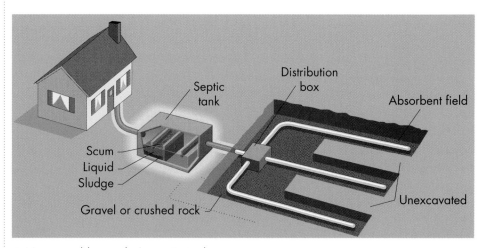

◆ Many rural homes have septic tanks.

remains in the septic tank must be pumped out of the tank every few years so that the tank does not fill up.

If properly maintained, septic tanks pose few environmental problems. However, NATURAL DISASTERS such as flooding can cause septic tanks to overflow and release their contents into the environment. This situation can be a problem if untreated sewage enters the groundwater supply or is carried to lakes and streams inhabited by organisms or used by people. Sewage released into the environment can also be harmful to soil quality, plants, and animals if it contains household materials such as detergents or cleaning fluids that are classified as HAZARDOUS WASTES. [*See also* LEACHING and SEWAGE TREATMENT PLANT.]

Sewage

▶The WASTEWATER flow that consists of liquid and semisolid wastes from homes, businesses, and SURFACE WATER. The wastewater is collected in a public sewer system that moves it along to a SEWAGE TREATMENT PLANT, where it is treated and released into rivers, lakes, or the sea.

Sewage is usually more than 95% water. It is classified according to amount, rate of flow, waste and BACTERIA it contains. Sewage is also classified as sanitary, commercial, industrial, or surface RUNOFF, depending on its source.

Sanitary sewage is waste material that originates in homes, hotels, apartment buildings, schools, hospitals, and prisons. It consists of human excrement, dirty water from personal washing and bathing, laundry and dishwashing water, animal and vegetable matter from food preparations, and other waste products from day-to-day living. Commercial sewage is produced by businesses such as stores, restaurants, service stations, offices, and laundromats. The makeup of commercial sewage is similar to that of sanitary sewage, although there may be more variation in the amounts of each kind of waste.

Industrial sewage, from factories that produce or assemble goods, is made up of some of the same wastes as those in sanitary and commercial sewage. However, this sewage may also include chemical products such as acids, oils, paints, and bleaches, among others. Surface runoff, also known as storm water, is PRECIPITATION that flows quickly over the land to definite passageways, such as street gutters that lead to **storm drains**. The precipitation—rain, sleet, or melted snow—absorbs gases and PARTICULATES from the air, filters through and flushes out vegetation and SOIL, and picks up oil and other chemicals dropped onto streets. The water then transports these pollutants to their collection point.

Sewer systems are designed for fast flow of materials from their sources to treatment facilities. The

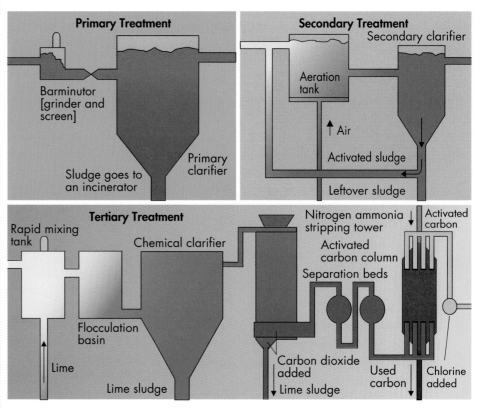

◆ Sewage goes through several steps at a treatment plant before it has been cleaned enough to be released into community waterways.

muck moist decaying matter; anything unclean or decaying.

raw sewage waste matter in its natural condition as carried off by a sewer, not yet treated or changed in any way.

storm drains underground basins to which gutters or pipes carry excess rainwater.

◆ California's South Lake Tahoe sewage facility is one of the few in the United States to provide a three-phase treatment. It includes screening, aerating, and the use of lime to make solid particles bunch together for disposal.

network of pipes has manholes, drainage inlets, flow-control mechanisms, and outlets from which the sewage enters a treatment plant.

TREATING SEWAGE

The Romans were the first to develop covered sewers. The drain-pipe that emptied sewage from the ancient Roman Forum is still in use today. From the Middle Ages (about 1450) until the mid-1800s, sewage was generally allowed to flow in open gutters through city streets and into waterways. However, this created a significant health hazard. The hazard was worse in cities that became crowded as a result of industrialization.

To enclose sewage, communities built stormwater drains, which carried sewage to nearby rivers. This soon led to severe WATER POLLUTION problems. To combat this problem, many city leaders looked for alternatives. One method, still practiced in some Asian countries, was to remove **raw sewage** from the water, treat it to kill the microorganisms, and to use the treated sewage as fertilizer.

Eventually engineers noted that natural waterways cleaned a fair amount of sewage by themselves. Sewage systems were designed to duplicate these natural conditions. The first method, called *trickling filtration,* sent sewage in specific amounts through a shallow tank filled with large stones. A more successful method, the *activated-sludge process,* passed compressed air through a shallow tank. The tank allowed for the DECOMPOSITION of the **muck** at the bottom by microorganisms.

PRIMARY, SECONDARY, AND TERTIARY TREATMENT

Today's treatment facilities may include these early sewage treatment methods. The most effective treatment of sewage involves three distinct steps—primary, secondary, and tertiary treatment. Not many facilities in the United States use all three methods, but some are working toward that goal.

Primary Treatment

Primary treatment removes about 30% of the organic wastes and bacteria from sewage. During primary treatment, sewage passes through a series of screens to remove large particles and debris. The water passing through the screen is sent into settling tanks. There the sewage may be mixed with a clotting material such as alum that makes some organic matter float to the water's surface, where it can be skimmed off.

Bacteria begin to break down material in the settling tank, using up the OXYGEN in the water as they do. The water is then released to another area. Sometimes, chlorine gas is added after primary treatment to kill many of the remaining bacteria.

◆ Liquid waste from city sewage treatment plants is put in vats to undergo further biological purification.

Secondary Treatment

Secondary treatment eliminates about 85% of the organic solids and oxygen-consuming waste that remain after primary treatment. Material from the primary settling tank flows into an aeration tank. In this tank, air is forced through the matter. The bacteria inside the tank move through the material, decomposing the organic substances to form less harmful matter. The liquid then flows into another tank, where sludge bacteria and some solids sink to the bottom as SLUDGE. The

sludge is pumped into a *sludge digestion tank* where bacteria break it down to produce METHANE, which can be burned as fuel to help power the treatment plant. Digested sludge can then be dried and buried, burned, or used as fertilizer.

Tertiary Treatment

Tertiary treatment produces an even purer discharge. This treatment may include chemical or radiation treatments. It may also involve the passage of the liquid through microscopic screens. This

final treatment makes the water safer for release into community waterways.

RURAL WASTE SYSTEMS

Many rural areas do not have public sewer systems. Approximately 25% of Americans use underground SEPTIC TANKS to collect and treat their sewage. These concrete or steel tanks, usually buried behind homes and other buildings, have pipes leading from the buildings to channel wastes.

Liquid wastes drain off into a *cesspool*—a series of canals that distribute it into sandy soil outside the tank. The solid sludge collects at the bottom of the tank, where it is attacked by bacteria and changed into gas and HUMUS. The gas escapes into the air through outlets. The humus is pumped out periodically and taken to treatment plants.

SEWAGE AND THE ENVIRONMENT

Although sewage treatment plants can produce methane, one of several ALTERNATIVE ENERGY SOURCES, they also create environmental problems. Sewage often contains large amounts of PHOSPHATES. These chemicals promote the growth of large populations of ALGAE—a situation called an ALGAL BLOOM. When the algae die, bacteria and other decay organisms absorb much of the DISSOLVED OXYGEN from the water. Too much oxygen depletion from the water threatens the survival of other water organisms.

Sometimes industrial sewage containing toxic and radioactive

◆ In primary treatment tanks (back), sewage is separated into solid and liquid parts after which they are exposed to aerobic decomposers in aeration tanks (front).

material is released into waterways. Toxic materials may pass through FOOD CHAINS and affect the lives of PLANTS, animals, and humans. [*See also* CARCINOGEN; CHLORINATION; CLEAN WATER ACT; COGENERATION; EFFLUENT; INDUSTRIAL WASTE TREATMENT; ORGANIC FARMING; TOXIC WASTE; and WASTEWATER, PRIMARY, SECONDARY, TERTIARY TREATMENT OF.]

Sewage Treatment Plant

▌▶ Facility where WASTEWATER is treated to remove pollutants. Sew-age treatment plants are designed to purify wastewater by the same processes that occur when organic wastes decompose in a lake or river. This purification process takes place in a series of stages.

PRIMARY TREATMENT

Primary treatment uses a series of screens to remove large debris, such as toothbrushes and jewelry, which are sometimes flushed down the toilet. Material not trapped in the screens proceeds to settling tanks. In this area, particles of organic matter settle to the bottom of the tank. Thus, the sewage is separated into liquid and solid components. The solid matter in sewage that settles to the bottom of settling tanks is known as SLUDGE.

SECONDARY TREATMENT

Secondary treatment starts with the products of primary treatment. Both the sludge and the fluid part of sewage is exposed to aerobic DE-COMPOSERS. The liquid from primary treatment is then run into an aeration tank. In the aeration tank, air is bubbled through the liquid to provide the aerobic decomposers with a HABITAT in which they can work. The liquid, which floats atop the sludge, is then piped into a second settling area. Chemicals that cause particles suspended in the liquid to clump together and settle out as sludge is added at this stage. The sludge that settles out of it is recycled into the aeration tank. Aerobic DECOMPOSITION reduces the volume of the sludge.

Another secondary treatment method is to spray the fluid from primary treatment onto trickling filters. Trickling filters are beds of gravel where dozens of kinds of aerobic decomposers live, including BACTERIA, FUNGI, protists, fly larvae, and worms. After passing through trickling filters, the fluid is carried to a settling tank, where solid matter settles out.

CHLORINATION

In most sewage treatment plants, following secondary treatment, the water is chlorinated. Chlorine kills any bacteria and ALGAE still living in the water. However, chlorine is also toxic to FISH. In addition, it can react with various organic molecules to form compounds that cause CANCER and birth defects in humans.

Chlorinated water that leaves a sewage treatment plant is released into rivers or RESERVOIRS. At this stage, there is not enough chlorine in most water to be a major threat to human health. But water supplies must be monitored to ensure that the concentration of chlorine compounds do not rise much further.

ADVANCED SEWAGE TREATMENT

Advanced sewage treatment is a series of processes designed to lower the amounts of specific pollutants left in water after primary and secondary treatment is completed. Specialized chemical or physical processes remove specific contaminants in different ways. For example, some contaminants are removed through SEDIMENTATION or DESALINIZATION.

The water left over after these advanced treatments are finished may be recycled for use in IRRIGATION. Most municipalities do not have the funds to establish advanced treatment plants. Such plants cost two times as much to build as primary and secondary treatment plants and four times as much to operate. Norway, Sweden, and Denmark have advanced sewage treatment facilities. Washington, DC, has cleaned up the Potomac River considerably using its advanced treatment plants. [*See also* ANAEROBIC; CLEAN WATER ACT; EFFLUENT; HEALTH AND DISEASE; HYDROLOGY; SAFE DRINKING WATER ACT; SEPTIC TANK; SEWAGE; and WATER QUALITY STANDARDS.]

Sick Building Syndrome

Condition that results from AIR POLLUTION contained inside closed structures such as office buildings, apartment houses, homes, and factories. Indoor air pollutants include chemicals, microorganisms, tobacco smoke, ASBESTOS, and RADON.

The Consumer Product Safety Commission and the National Institute for Occupational Safety and Health (NIOSH) have discovered that the air inside a home or office can contain hundreds of chemicals that are potentially harmful to people. New walls, doors, carpeting, flooring, counters, paints, and varnishes can release chemical gases into the ENVIRONMENT for months, or even years, after they are installed. Cleaning products, air fresheners, cosmetics, fumes from gas pilot lights, gases and energy given off by televisions, copiers, computers, and stereos, dry-cleaned clothing, and PESTICIDES all add to the chemical mix contained in indoor air.

TYPES OF INDOOR POLLUTANTS

One of the most significant indoor chemical pollutants is formaldehyde. It is used to manufacture building materials, such as paneling, plywood, cabinets, furniture, ceiling panels, no-iron fabrics, wallpaper, paper towels, and carpeting. It is also used in the manufacture of personal care products, such as soaps, shampoos, deodorants, and toothpastes. Formaldehyde fumes released from products can build up in a closed building. These fumes can irritate the eyes, nose, and throat.

Some medical researchers believe formaldehyde is a sensitizer—a chemical that makes a person sensitive to other chemicals in the environment. This condition is called *multiple chemical sensitivity* (MCS). MCS symptoms include rashes, stomach upsets, headaches, and breathing problems in people who are exposed to chemicals such as NATURAL GAS, AUTOMOBILE exhaust, tobacco smoke, alcohol, dust, perfumes, PLASTIC, and even many kinds of food.

Organisms that enter buildings, such as BACTERIA, FUNGI, and dust mites, thrive and multiply in moist areas such as bathrooms, kitchens, and within ventilation systems. VIRUSES may also spread in the air or water inside buildings. Many organisms in air or water can cause allergies. Some cause respiratory infections, such as influenza, tuberculosis, and bronchial asthma. Among the most dangerous microorganisms are the bacteria that thrive in ventilation systems and cause Legionnaire's disease. Legionnaire's disease is a serious kind of pneumonia.

According to the ENVIRONMENTAL PROTECTION AGENCY (EPA), tobacco smoke inside buildings can cause ear, lung, and throat problems and can aggravate asthma. This is especially a problem for young children. Tobacco smoke contains more than 4,000 chemicals. About 43 of these chemicals are believed to be CARCINOGENS. Most are present in greater amounts in smoke coming from the end of the burning cig-

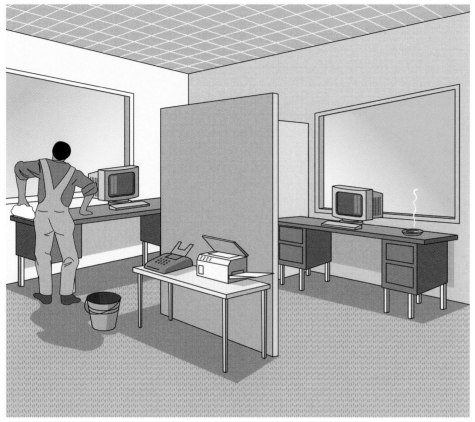

◆ Many sources of air pollution can be found inside a regular office.

arette than in the smoke inhaled by the smoker. The EPA reports that 3,000 nonsmokers per year die from lung CANCER caused by tobacco smoke. To help solve this problem, many private and public building managers have banned smoking inside their buildings.

Asbestos was once commonly used in insulation and in building materials, such as paint, plaster, and ceiling and wall tiles. Asbestos was suited to these uses because of its strength, flexibility, and fire resistance. However, asbestos fibers can break away from these materials and be carried in the air. In the 1960s, it was discovered that these fibers can cause lung cancer and other lung diseases.

Since the early 1970s, strict EPA regulations have banned the use of asbestos in building materials. Laws passed in the 1980s called for the removal of asbestos from public buildings. However, it is now believed that it is safer to make certain that materials that have asbestos be contained with a binder or sealer. This practice is recommended because removing asbestos can release a shower of asbestos fibers into the air, where they can be easily inhaled.

Radon is a radioactive gas that is given off by rocks and SOIL. Radon seeps into buildings through pipes, drains, floors, and pumps. According to the EPA, radon causes about 14,000 deaths from lung cancer each year. These statistics make radon the second leading cause of lung cancer.

The EPA considers indoor air pollution one of the top five environmental health problems. Such pollution is a greater health risk than TOXIC WASTE dumps, factory emissions, and PESTICIDES. To help address the problem, the EPA increased its budget for dealing with indoor air pollution from $35,000 in 1987 to $7 million in 1993.

AVOIDING INDOOR POLLUTANTS

Ventilation systems are the best solution to the problem of indoor air pollution, particularly in energy-efficient office buildings that are fitted with windows that cannot be opened. Such buildings have ventilation systems, but building managers may fail to mix enough fresh outside air into the system because it costs money to heat or cool that air. There are no regulations that require building managers to provide proper ventilation in office buildings. However, agencies such as NIOSH are looking to Congress to write standards and regulations for indoor air quality.

Some scientists recommend that homeowners become aware of the many chemical fumes that may be emitted from materials in their homes, especially new materials, and that they install and properly maintain good ventilation systems. Good cleaning practices, in addition to the use of dehumidifiers, to control indoor moisture, can keep the home free of many air pollutants. [See also CARBON DIOXIDE; CARBON MONOXIDE; HEALTH AND DISEASE; RADIOACTIVITY; and RADON.]

Silent Spring

See CARSON, RACHEL LOUISE

Silt

See SOIL

Silviculture

▶ The care and cultivation of trees. Foresters who practice silviculture are sometimes called *silviculturalists*. They are trained at FORESTRY schools in the techniques of "silvics," which is forest management. Silviculturalists carry out scientific observations, research, and experiments to learn about forest growth.

Before the timber industry employed silviculture, unprofitable trees were cut down to give profitable trees more room to grow. Often, unprofitable trees were killed by "girdling," a method in which V-shaped notches are cut into them and filled with poison. Trees were routinely pruned to direct growth into the main trunk—which was the valuable wood—and away from unprofitable branches. In many cases, PLANTS on the forest floor were removed because they were thought to steal resources from the trees.

Today, many silviculturalists favor what is called *selective cut-*

◆ Silviculture is the care and cultivation of forest trees.

ting. In this practice, trees of different SPECIES and ages are selectively cut. The trees are removed individually or in small groups, while the rest of the forest is not disturbed.

Gradually, new seedlings replace the cut trees. Selective cutting is much less damaging to forest ECOSYSTEMS than any other method of tree harvesting.

Silviculture is slowly but surely being transformed into an ethic of using our forest resources while preserving the integrity of forest ecosystems. The term *silviculture* is being redefined, and its practice continues to be reoriented toward CONSERVATION. [*See also* AGROFORESTRY; CLEAR-CUTTING; DEFORESTATION; FOREST; FOREST PRODUCTS INDUSTRY; and SUSTAINABLE DEVELOPMENT.]

Sludge

▶A thick liquid waste material containing BACTERIA, FUNGI, organic matter, and MINERALS. Sludge is a waste material that is a concentrated suspension of solids in a liquid. Usually, the liquid is water.

Sludge is formed in various manufacturing processes and SEWAGE TREATMENT PLANTS. Sludge often contains toxic substances and must be treated and disposed of safely by making use of sludge treatment and disposal systems.

In a water treatment plant, water that is to be used for drinking, cooking, and bathing is treated with materials that destroy living organisms and filter out debris and unwanted substances. These unwanted materials are usually separated from water by allowing them to settle to the bottom of the treatment tank as sludge. [*See also* SEDIMENTATION; SEPTIC TANK; WASTE MANAGEMENT; WASTEWATER; WATER, DRINKING; WATER QUALITY STANDARDS; and WATER TREATMENT.]

Smelter

▶A furnace for melting mineral ores to isolate the metals they contain. An ore is a naturally occurring mineral or mixture of minerals. Most of the minerals used by people are dug out of the ground in the form of ores. The ores are then processed to purify them or to extract a particular substance.

Metals are extracted from their ores by smelting, or the process of heating the ore until the metal melts. Many useful substances, both metals and nonmetals, are extracted from their ores by heat. Examples of such substances include silicon, COPPER, tin, and LEAD.

The nonmetallic material removed by smelting is called *slag*. It consists mostly of ash and mixed oxides of silicon, sulfur, phosphorus, and ALUMINUM. Smelter slag can be a pollutant when not properly disposed of.

Copper was probably the first metal extracted from ore by smelting. The ancient Sumerians, in what is today southern Iran, smelted copper ores with wood fires. The ores were not purified, and sometimes a mixture containing both tin and copper was smelted. Heating such a mixture produces bronze, an alloy of tin and copper that is hard and resists corrosion. Bronze objects from nearly 5,000 years ago have been found in Egypt. [*See also* MINING.]

◆ Copper is extracted from its ore by heating the ore until the metal melts.

Smog

▶Irritating haze resulting from the sun's effect on certain pollutants in the air, notably those from AUTOMOBILE exhaust. Smog also refers to a mixture of smoke and fog. There are two main types of smog: photochemical smog and industrial smog.

PHOTOCHEMICAL SMOG

Photochemical smog is a mixture of primary pollutants and secondary pollutants. A primary pollutant is one that is introduced into the air directly, such as CARBON MONOXIDE, HYDROCARBONS, and NITROGEN DIOXIDE. A secondary pollutant is exhaust that is formed by the chemical reactions between primary pollutants.

In photochemical smog, the primary pollutants are hydrocarbons and nitrogen oxides. The main source of these pollutants is from motor vehicles. Chemicals in motor vehicle emissions react with OXYGEN in the air and with sunlight to produce OZONE, peroxyacetyl nitrates (PAN), aldehydes, and other compounds.

The pollutants that cause photochemical smog directly affect lungs and eyes, causing irritation in these organs. They can cause changes in lung functions and are harmful to people with already-existing lung disease. These pollutants are extremely toxic to plants and can also physically weaken such materials as rubber and fabrics.

INDUSTRIAL SMOG

Industrial smog is a mixture of SULFUR DIOXIDE and a variety of microscopic solid and liquid particles suspended in air. This type of smog comes more from stationary sources (power plants, furnaces, incinerators, etc.) than from motor vehicles. Industrial smog is sometimes called grey-air smog. Due to our relatively strong AIR POLLUTION laws, industrial smog is kept more in check in the United States than it is in countries that have no air pollution control laws.

Sulfur oxides, in combination with water and oxygen, can turn into sulfuric acid in the ATMOSPHERE. This acid falls to Earth as ACID RAIN. Acid rain can dissolve marble and eat away iron and steel. In humans, sulfur oxides can affect the respiratory system by damaging the cells of the breathing passages and of the lungs. PARTICULATES that enter the respiratory tract can do damage to the respiratory system. In combination with acid, the damage caused by particulates is magnified. This type of smog is also very dangerous for people who already have lung or heart conditions.

THERMAL OR TEMPERATURE INVERSIONS

If smog is very concentrated, it can kill people. It can become very concentrated by a thermal inversion. Normally, hot air rises. As it rises, it takes POLLUTION up and away from Earth's surface. In a thermal inversion, cool air is trapped by a layer of warm air, preventing ground-level air from rising and taking away pollutants. As a result pollution builds up near Earth's surface. Inversions are particularly dangerous when they occur in cities that are located in a valley or a basin, because the air is trapped and wind cannot remove the smog. [*See also* CLEAN AIR ACT.]

Softwood

▶Wood with a certain type of vascular tissue, commonly used in building construction. Softwoods are also used to make household furnishings, shelving, and some kinds of small boats and canoes. Pine is a popular softwood.

Softwood trees do not necessarily have wood that is softer than

THE LANGUAGE OF THE ENVIRONMENT

coniferous trees trees whose seeds form in cones; most coniferous trees do not shed their leaves all at the same time.

deciduous trees trees that shed all their leaves each year.

hardwoods. Trees are considered to be softwood or hardwood mainly on the basis of the type of xylem they have. Xylem is vascular tissue that carries water and MINERALS throughout a tree.

The xylem in softwoods has rays that are narrower and shorter than those in hardwoods. Rays are structures that pass nutrients to other parts of the xylem and also store starch and lipids that help nourish the tree. Softwoods contain rays that are usually 1 cell wide and between 1 and 20 cells high. By contrast, the rays of hardwood trees are 1 to 20 cells wide and can be several hundred cells high.

Most **coniferous trees** are softwoods. Most **deciduous trees** are hardwoods. However, there are exceptions. Hemlock, for example, is a coniferous tree that has xylem, placing it in the hardwood category. [*See also* CONIFEROUS FOREST.]

Soil

▌The very thin layer of natural materials on the surface of Earth containing both organic and inorganic materials capable of supporting life. It may be covered with water, such as at the bottom of the sea, or it may be the land surface. On land, it is located between the ATMOSPHERE and the deepest roots of PLANTS. The word "soil" comes from the Latin *solum,* which means the "floor" or "ground."

Soils have three components—solid, liquid, and gas. The solid component of soil consists of MINERALS and organic particles. More than 90% of Earth's land surface is made of the following eight elements: OXYGEN, silicon, ALUMINUM, iron, calcium, sodium, potassium, and magnesium.

These elements combine together and with other elements to form more than 2,000 different types of minerals found on Earth. Examples of the most common minerals are silicon oxide, aluminum oxide, iron oxide, calcium oxide, and magnesium oxide. The first three make up 87% of the ocean floor. All of these minerals make up 94% of the land surface. On land, the liquid component of soil, which is water from rain, and the gas component of soil, which consists mostly of nitrogen, oxygen and carbon dioxide, are found in a network of pores that separate the mineral particles.

WEATHERING

Soil is formed through the process of WEATHERING of rocks. There are three groups of rocks. Igneous rocks are formed from the molten material thrown up to the surface of the land through volcanic action or from underground volcanic activity. Sedimentary rocks are formed from sediment which is made up of mineral and organic material in suspension in the air or water. Sediment is constantly being moved from one site to another. Over long periods of time, it can become solid rock. Metamorphic rocks are made when igneous or sedimentary rock are reformed by very high temperatures and pressures.

There are different kinds of forces involved in the weathering process of rocks. These include the action of wind and water, ice expansion, mechanical grinding, plant root growth, and temperature changes. In addition to weathering, other factors that affect soil development are *topography* (the form of a surface, which then plays a part in water movement and EROSION), time, the nature of the rocks from which the soil is being made, and organisms that live in the soil. Fragments produced by weathering are changed by mechanical and chemical processes and mixed with organic materials. The organic matter in soil consists of animal wastes and animals and plants in various stages of decay. Biological processes occur, and finally, soil is formed.

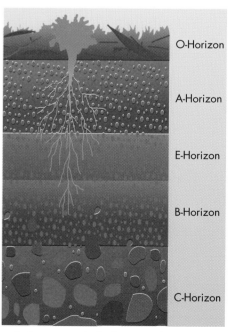

O-Horizon

A-Horizon

E-Horizon

B-Horizon

C-Horizon

◆ Soil is composed of various layers, including the O horizon, or topmost layer, and the B horizon, or subsoil.

LAYERS OF SOIL

Soil is made up of several distinct layers. The top horizon, or layer, is called the O horizon. This layer is composed of surface litter made up of fallen leaves and other organic material. The A horizon, below the O horizon, consists of TOPSOIL, which is made up of partially decomposed organic matter, living organisms, and some inorganic materials. The E horizon, or zone of eluviation, is the area where dissolved or suspended materials are brought downward by descending rainwater in the process called LEACHING. The B horizon underneath the E horizon is the subsoil, where mineral compounds and material has leached down. Parent rocks, the rock from which the soil is made, lie in the C horizon, under the subsoil. Below this is the R horizon, called the *bedrock*.

ORGANISMS AND ORGANIC COMPOUNDS IN SOIL

Soils support plant life. Most plants get nutrients from the soil through their roots. These plants are eaten by HERBIVORES. There are the large herbivores found above ground, but the soil also contains herbivores. Some examples of soil herbivores are sowbugs, millipedes, springtails, ants, beetles, grubs, snails, and slugs. Small MAMMALS, such as squirrels, gophers, woodchucks, and mice, are also common soil herbivores.

The herbivores are eaten by CARNIVORES, such as centipedes, beetles, moles, shrews, spiders, and mites. When plants and animals die, their remains are decomposed by such organisms as BACTERIA, FUNGI, yeasts and molds, and earthworms, which live in the two top layers of most soils. The end products of DECOMPOSITION are recycled into the organic matter of soil.

The organic compound of most soil is very small. It makes up from 2% to 5% of the soil and is made mostly from decaying plant material. The rest is from animal waste and decaying bodies. Brown, decaying material in the soil is called humus. Organic materials have important functions in the soil. They hold minerals together, serve as nutrient sources for plants, provide food for microorganisms, and store moisture.

EROSION

Soil erosion is the wearing away and the transportation of soil from one place to another. This is done by the action of wind and water. The forces of erosion may remove the topsoil entirely. An area in which erosion occurs faster than weathering occurs is in danger of losing its topsoil, making it unable to support life. It has been estimated that approximately 1 to 5 tons of soil per acre is eroded every year under natural conditions. It is also estimated that the same amount per acre is formed by natural processes per year. On cropland, more than 8 tons of soil can be lost per acre per year from erosion.

Aside from loss of topsoil, erosion decreases the capacity of the soil to hold water and decreases the depth of the root zone. Erosion also causes GULLYING, the formation of gullies as water flows along the natural depressions in the soil. The displaced soil pollutes lakes and

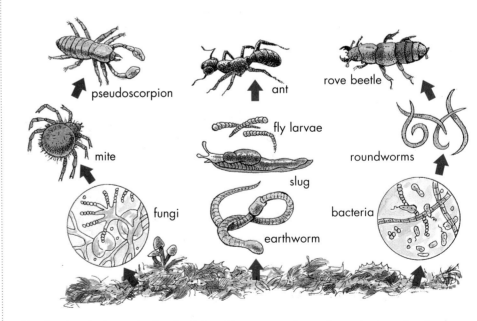

◆ This food web shows the food relationships among the various plants and animals in soil.

◆ Earthworms function as decomposers of dead plants and animals in soil.

eventually fills in lakes. Several methods have been used to slow soil erosion. These include maintaining a ground cover of plants, planting grass along waterways, and planting trees to break the force of winds. CONTOUR FARMING and CROP ROTATION are farming methods that help prevent erosion.

CLASSIFICATION OF SOIL: TEXTURAL CLASSES AND OTHER CLASSES

Any soil can be categorized according to its texture. This is done by analyzing varying proportions of sand, silt, and clay, as well as organic matter present in the soil. There are three kinds of rock particles present in soil—sand, silt, and clay. Sand particles are relatively large and irregularly shaped with large pores between them. Water and air readily flow through sand. Silt particles are smaller than sand, and thus can hold water better than

sand. Clay particles are the smallest of the three and can hold water very tightly. There are almost no large spaces found in clay particles, but innumerable tiny spaces.

Knowing the textural class of a particular soil will help to determine what kind of vegetation the soil will support. Clay loams are ideal for horticulture or gardening. Loam soils with a lower percent of clay or silt are still good but will not hold water as well. Corn will grow well on clay soils, but potatoes will not.

Other classification systems exist and may be used to predict the productivity and potential of soils, as does the soil texture system. The most recent soil classification system, published by the U.S. DEPARTMENT OF AGRICULTURE and the land-grant colleges, classifies soils based on such characteristics as quality and quantity of organic matter in the surface layers and the availability of moisture.

TROPICAL SOIL AND TEMPERATE SOIL

One way to classify soil is based on where it are found—in tropical soil and temperate regions. The primary differences between tropical and temperate soil are the result of differences in CLIMATE. Winter temperatures of the temperate zones are a factor in retarding the weathering process and biological action; and so soil development is slower in temperate zones than it is in the tropics. Weathering in the tropics can proceed as much as four times faster than in temperate zones.

The millions of years of leaching of soil in the tropics have led to the destruction of most original materials. What remains is a very high proportion of iron or aluminum. Iron- and aluminum-rich soils, called *oxisols,* are found only in the tropics. When oxisols are exposed to sunlight and water, they turn to rock. In tropical agriculture, therefore, it is important that covering vegetation not be removed to prevent the soil underneath it from turning to rock. This is a problem in developing countries, where providing food for a rapidly growing population is of utmost concern.

THE EFFECTS OF HUMAN ACTIVITY ON SOIL

In addition to natural processes, human activities interfere with nutrient cycling. Humans harvest crops that would have been eaten by herbivores and would have been part of the food chain and biogeochemical processes. The complete removal of crops from the soil makes it less fertile because nutrients are taken out of it.

Soil Conservation

◗The protection of SOIL from EROSION and loss of fertility. Soil conservation was first begun in the United States when Congress passed the Soil Conservation Act of 1935. This action was taken in response to the dust storms of the 1930s, which blew away many thousands of tons of rich TOPSOIL in the Midwest.

In 1937, at the suggestion of President Franklin Delano ROOSEVELT, the states organized soil-conservation districts. Today, there are 3,000 of these districts, which are governed by a five-person board. With the guidance of the Soil Conservation Service, which is part of the DEPARTMENT OF AGRICULTURE, farmers work to devise soil-conservation plans that are best for their areas.

CONSERVATION PLANNING

The first step in devising a soil-conservation plan is to survey the land to determine the best use for it. The soil's fertility is measured through studies of soil particle size, mineral content, and microbial activity. Soil depth, drainage, and the slope of the land are also examined.

Based on these factors, the land is rated in one of eight classes. Classes 1 through 4 are suitable for growing crops. Class 5 is considered of some value for growing crops despite such qualities as stoniness or being too wet or too

dry. Classes 6 through 8 are unsuitable for growing crops and are considered best for use as GRAZING land, FORESTS, WILDLIFE HABITATS, or MINING areas. Once the best use of the land has been determined, a CONSERVATION plan can be developed.

CONSERVATION METHODS FOR RANGELAND AND FORESTS

To conserve the soil on RANGELAND, animals cannot be allowed to destroy PLANTS by eating them down to the ground. The animals must be moved from field to field to give the plants time to grow back.

Forests must be protected from fire. Any logging must be done selectively so that parts of a forest are not left bare of trees, which exposes the soil to erosion. Planting new trees also helps keep forests intact.

CONSERVATION METHODS FOR FARMLAND

Soil conservation on farmland is challenging because the farmer must disturb the soil to plant crops. Growing plants also removes nutrients from the soil. The objective is to farm the land so that it will be as productive as possible without harming the soil. Some of the most important soil-conservation methods for farmland are NO-TILL AGRICULTURE, CROP ROTATION, CONTOUR FARMING, and fertility maintenance.

No-Till Agriculture

In no-till agriculture, plant debris is left on the fields in the fall. In the spring, only the digging needed to plant the new crop is done. The debris adds nutrients to the soil and protects it against erosion by wind and rain. However, because the debris fosters the growth of weeds and INSECTS, methods of PEST CONTROL must be used.

Crop Rotation

Crop rotation is the alternating of crops from year to year. For example, wheat might be planted the first year, a leafy plant such as lettuce the second year, and a LEGUME such as peas the third year. Crop rotation helps restore nutrients to the soil. It has the added advantage of helping to keep insect pests under control. Each type of insect

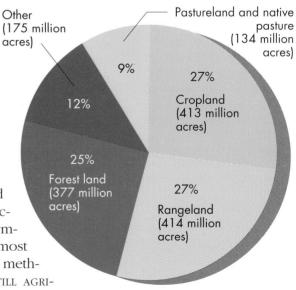

◆ This chart shows how land is used in the United States.

Other (175 million acres) 9%

Pastureland and native pasture (134 million acres) 27%

Cropland (413 million acres) 27%

12%

Forest land (377 million acres) 25%

Rangeland (414 million acres) 27%

tends to prefer a certain kind of plant. Rotating crops prevents the buildup of insect populations year after year.

Contour Farming

In contour farming, small furrows are plowed parallel to small slopes. The furrows act as dams to stop the downhill flow of water and soil.

In TERRACING, a variation of contour farming, flat areas are built up in a stair-step fashion. The flat area holds water so that it soaks into the soil. Another variation is to alternate rows of one crop with rows of a denser-growing crop. This is called *strip-cropping*. The crop that grows more densely serves as a windbreak and also protects the soil from water erosion.

Maintaining Fertility

Synthetic fertilizers supply minerals. Soil also needs organic matter, both because it provides nutrients and because it improves the texture of the soil, making it clump. Soil that clumps does not easily erode. Animal feces, or manure, is one source of organic matter. So-called green manure is another. Green-manure plants may be planted and plowed under just to add organic matter to the soil. LEGUMES, pod-bearing plants such as beans and peas, are common green-manure plants. These plants not only provide a crop but also add nitrogen and organic matter to the soil. [*See also* AGROECOLOGY; BIOGEOCHEMICAL CYCLE; CHEMICAL CYCLES; DECOMPOSITION; HUMUS; NITROGEN FIXING; and ORGANIC FARMING.]

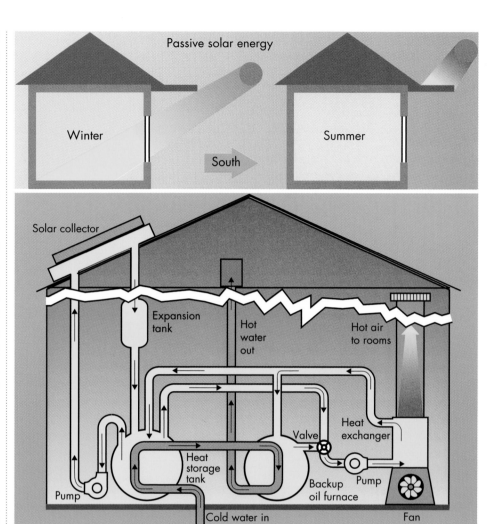

◆ The solar energy collected on the roof heats the water running through the pipes. The heated water is stored in the cellar and is used in various ways.

Solar Energy

▌Renewable form of energy that results from harnessing the energy in sunlight. Solar energy can provide far more energy than all the FOSSIL FUELS on Earth. In fact, many scientists believe that harnessing and using the sun's energy will reduce the use of fossil fuels and the AIR POLLUTION and environmental damage associated with it.

Many environmentalists recommend an increased use of solar enegy because it does not contribute to air pollution or WATER POLLUTION. Today, about 1% of the world's energy supply comes from solar energy. However, the use of solar energy is steadily growing. Scientists believe solar energy will someday replace fossil fuels as the main source of ELECTRICITY. However, before this occurs, methods for lowering the costs of harnessing this energy source needs to be developed.

HARNESSING THE SUN'S ENERGY

Four methods are currently used to harness the sun's energy. In general, these methods involve ways of using solar energy to heat buildings and to generate electricity.

Passive Solar Heating

Passive SOLAR HEATING is the simplest and most inexpensive way to use the sun's energy. Passive solar energy is used to heat homes and small buildings without the use of pumps or fans.

The process of passive solar heating is similar to the way the inside of a car heats up on a hot summer day. Buildings located in the Northern Hemisphere that are designed for passive solar heating have energy-efficient, double- or triple-paned windows that face south. South-facing windows can absorb the greatest amount of the sun's heat during the day. Such buildings are also built with thick layers of stone or concrete that take in the heat of the sun and release it slowly back to the ENVIRONMENT. In addition, thick insulation and the use of heavy drapes over the windows during nighttime hours help reduce heat loss from a building during nondaylight periods. An efficient passive solar heating system can heat a building even in very cold WEATHER without the use of fans or pumps to distribute the heat. The only requirement is ample amounts of sunlight.

Active Solar Heating

Active solar heating takes the collection and use of the sun's radiant energy a step further than passive solar heating. In this method, a series of solar collectors are used to absorb solar energy and convert it into heat. Buildings equipped with active solar heating systems typically have several solar collectors mounted on the roof or the south sides of the building. The collectors are enclosed in a dark, insulated box that absorbs the sun's energy. Beneath the collectors are a series of water-filled tubes that take in the heat collected. This heated water is then pumped to radiators around the building or to hot-water storage tanks. As the heat energy is released from the water to warm the surroundings, the water moves back to the collectors to be reheated.

Photovoltaic Cells

The third method for capturing the sun's energy makes use of solar cells, or photovoltaics. PHOTOVOLTAIC CELLS convert sunlight directly into electricity. These cells were created in 1954 at Bell Laboratories. Today, they are commonly used to power cars, calculators, watches, and satellites. Small villages in India and many homes in the United States also use electricity generated by photovoltaic cells.

Photovoltaic cells use purified silicon (from sand) and other chemicals that conduct ELECTRICITY when struck by sunlight. Since solar cells are small, they must be linked together in large sheets to provide enough electricity for a building. Some scientists estimate that photovoltaics could supply as much as 25% of the world's energy needs within 50 years. Currently, however, solar cells are expensive and not cost-effective compared to the use of fossil fuels.

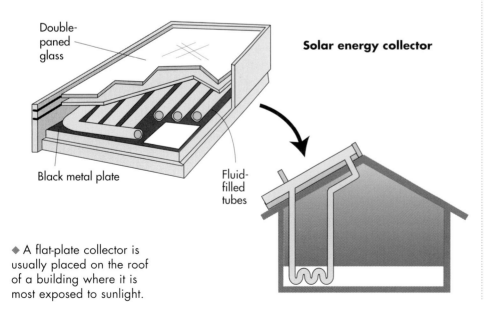

Double-paned glass

Black metal plate

Fluid-filled tubes

Solar energy collector

◆ A flat-plate collector is usually placed on the roof of a building where it is most exposed to sunlight.

Solar Heating

◆ With active solar heating, heat from solar collectors is circulated through pipes to radiators and hot-water storage tanks.

▶The use of the radiant energy of the sun to heat buildings. The radiant energy given off by the sun is the source of almost all energy on Earth. This energy, especially that in the form of light and heat, makes it possible for living things to survive on Earth. SOLAR ENERGY is also the driving force behind the WATER CYCLE and is responsible for the winds and OCEAN CURRENTS that control CLIMATES throughout the world. People, too, make use of solar energy to provide themselves with light, heat, and ELECTRICITY.

THE HEATING EFFECTS OF THE SUN

To some degree, people have always made use of solar energy as a source of heat. Much of the solar energy that reaches Earth is in the form of visible light and infrared RADIATION. Both forms of energy heat Earth. For example, much of the visible light that reaches Earth is absorbed by objects on Earth's surface. As substances absorb light energy, the energy is changed to heat, or thermal energy. The sun also gives off infrared radiation. Gases in the ATMOSPHERE, such as CARBON DIOXIDE and METHANE, allow visible light to pass through them, but they trap infrared radiation. The gases that carry out this task are called GREENHOUSE GASES because, like the glass in a greenhouse, they trap infrared radiation and change it to heat within the atmosphere. This trapped heat energy helps regulate temperatures on Earth.

Solar-thermal Power Plants

Another promising technique for generating electricity from sunlight is the use of solar-thermal power plants. These power plants use large mirrors to reflect sunlight onto a series of solar collectors containing water- or oil-filled tubes. The heat absorbed by the fluid in the tubes is used to generate electricity by turning the blades of a turbine connected to a generator. In California's Mojave Desert, one solar-thermal power plant generates 274 megawatts of electricity. This is enough to meet the demands of a small city. It is currently used to supplement the power needs of Los Angeles.

Like other solar energy systems, solar-thermal power plants do not produce water, air, or land pollution. The cost for generating electricity in this way is similar to the use of fossil fuels and NUCLEAR POWER. A drawback to the system is that the plants require a great deal of space for their mirrors. The mirrors must be cleaned frequently to be most effective and must be placed in regions that receive a great deal of sunlight.

SOLAR ENERGY AND THE FUTURE

Solar energy is an example of a RENEWABLE RESOURCE. Solar energy is inexhaustible, even if it is continually used. Many scientists consider solar energy to be the ultimate source of energy and the answer to the world's future energy problems. [*See also* ALTERNATIVE ENERGY SOURCES; ENERGY EFFICIENCY; NONRENEWABLE RESOURCES; and TELKES, MARIA.]

People in many parts of the world are using the energy of sunlight to provide heat for their homes. The use of solar heating has many advantages over other heating methods. One main advantage is that solar heating makes use of a renewable NATURAL RESOURCE that is in abundant supply. In addition, solar energy, unlike FOSSIL FUELS and NUCLEAR POWER, is nonpolluting. Another advantage to the use of solar energy is that it is free to the consumer when it is used passively. The main disadvantage to the use of solar energy as a heating source is that the sun's energy is available only during daylight hours on clear days. In addition, the times during which the sun's energy is available varies greatly with the seasons, according to where on Earth a building is located.

PASSIVE SOLAR HEATING

Passive solar heating systems make direct use of the sun's energy without the aid of special devices. A home that makes use of passive solar heating has many windows that face toward the south. South-facing windows are able to take in the sun's energy for the greatest number of hours in a day. In addition, homes that are designed to make use of passive solar heating are usually constructed from dark materials with rough surfaces. Such materials tend to absorb rather than reflect sunlight and to change this light energy to heat.

Buildings that make use of passive solar energy often have large, glass-enclosed areas that resemble greenhouses. Like a greenhouse, these areas allow sunlight to pass through the glass. Once inside, the light is absorbed by materials making up the floors and walls of the room and changed to heat that is released back into the room.

ACTIVE SOLAR ENERGY

An active solar heating system combines devices such as tubes, tanks, fluids, and pumps with the elements of a passive solar heating system. The devices in an active solar heating system help collect, store, and circulate the heat obtained from the sun. Because of these devices, an active solar heating system is better able to store and distribute heat energy.

The main component of most active solar heating systems is the solar collector. A solar collector is a device that is mounted outside the building, usually on the roof, where it is exposed to the sun's energy. The most commonly used solar collector is a flat-plate collector. This device is a box that is made up of a black metal plate at its bottom, a network of fluid-filled tubes that are usually connected to a water collection device, and a double-paned layer of glass.

As sunlight strikes the flat-plate collector, the light is absorbed by the black metal at the bottom of the collector box. Here, the absorbed light is changed to heat that is released into the collector box. The heat in the collector box causes the fluid inside the network of tubes to become warmer. The heated fluid is then pumped to the water collection device, where it heats the water in the container. The heated water may then flow through a closed system of tubes running throughout the home, giving up its heat to the rooms, or being used as a source of hot water. After the system of tubes or the water collection device gives up its heat, the cooler water is pumped back to the solar collector. Once in the solar collector, the fluid is again heated and the cycle begins again as this heated water is pumped back to the water collection device.

Unlike a passive solar heating system, an active system requires some electricity to operate. The electricity powers the pumps that move the heated fluid through the system. Although it requires some electrical energy to operate, active solar energy systems save FOSSIL FUELS since they do not burn fossil fuels directly and require less electrical energy than other heating systems, such as heat pumps and electric heat. [*See also* ALTERNATIVE ENERGY SOURCES; RENEWABLE RESOURCES; and TELKES, MARIA.]

Solid Waste

▶ Waste from human activities usually disposed of in LANDFILLS. Solid waste differs from liquid and gas wastes, which are treated and disposed of in different ways. Solid wastes include wastes from households and businesses, such as appliances, food, or metal scraps (often called municipal waste, trash, or GARBAGE), factory wastes, debris from lawns and gardens, and wastes from the demolition of old buildings and the construction of new buildings (construction waste). The sources of U.S. solid waste are MINING and oil and gas production (75%), agriculture (13%) industry (9.5%), municipal (1.5%), and sewage SLUDGE (1%).

Solid wastes include materials that can be reused or recycled, such as bottles, cans, and paper, as well as other materials that may be hazardous. Solid wastes are considered problems because they may contain hazardous materials and because disposing of materials on land is damaging the ENVIRONMENT and becoming increasingly expensive.

SOLID WASTE DISPOSAL

Disposing of solid waste on land has been practiced for centuries. However, as the human population (and the amount of waste) increased, cities turned to more economical methods of waste disposal. In the 1930s, cities began constructing sanitary landfills in which solid waste was compressed by heavy machinery and covered with SOIL each day. This decreased the potential spread of disease and the growth of populations of nuisance SPECIES like rats and mice.

In sanitary landfills, however, wastes are not separated nor do they protect certain parts of the environment. For example, compressing and covering solid waste concentrates the waste, which can reduce or eliminate its DECOMPOSITION. A professor at the University of Arizona found twenty-year-old newspapers and undecayed food in modern landfills. Additionally, water seeping through landfills can move into groundwater and carry hazardous chemicals into drinking water supplies.

REGULATING DISPOSAL

Recently, new rules and regulations in the United States have limited the kinds of materials that can be placed in landfills. Under the RESOURCE CONSERVATION AND RECOVERY ACT (RCRA) of 1976, hazardous materials are no longer allowed in local landfills. Materials like grass and leaves that decay naturally are also banned. Strict rules now govern the construction of landfills so that groundwater is protected.

Water and waste mix together to produce an acidic liquid called *leachate*. Leachate can contain a mix of hazardous materials that contaminate soil and groundwater. Modern landfills need PLASTIC and clay liners and caps to reduce the movement of water into the waste.

Many communities have found that operating landfills is expensive. Some of the first attempts to reduce the amount of solid waste used open burning or incinerators. However, both open burning and faulty incinerators produced AIR POLLUTION. Some communities found ways to use incinerators to generate ELECTRICITY in a way similar to burning COAL or oil. These incinerators produced other problems, including toxic ash that had to be disposed of as a HAZARDOUS WASTE. In addition, much of the material in solid waste does not burn, so no heat is produced from trying to burn glass or metal. As a result of these problems, many waste incinerators have been closed.

◆ The landfill at Fresh Kills, Staten Island, New York, is claimed to be the largest in the world, according to the New York City Department of Sanitation.

ALTERNATIVE TO LANDFILLS

Recycling programs have reduced the amount of material disposed of in landfills. Recycling removes many kinds of materials that used to be thought of as trash from disposed wastes. Often, community recycling reduces the amount of solid waste placed in a landfill by 30%. Some communities have set a recycling target of 50%. Most paper, glass jars and bottles, metal cans, and many plastics can be recycled. Some scientists have suggested that as much as 90% of solid waste can be recovered for reuse or recycling. Some communities and countries have attacked the solid waste problem by banning certain kinds of package materials to keep these materials from becoming solid waste. [*See also* POLLUTION; RECYCLING, REDUCING, REUSING; SOLID WASTE DISPOSAL ACT; and WASTE REDUCTION.]

packaged and partly because POPULATION GROWTH had resulted in more people producing waste. More and more people were living in cities, which meant that large amounts of waste were quickly building up near metropolitan areas. This buildup led to ugly landscapes, public health hazards, AIR POLLUTION, and WATER POLLUTION.

RESULTS OF THE ACT

The purpose of the Solid Waste Disposal Act was to find ways to deal with the disposal of solid waste. The act supported research and education programs about solid waste, and it prompted the recycling of materials. Under the act, state governments were directed to manage their own solid waste problems. The act provided money for them to create plans for managing waste.

By 1975, every state had its own plan for solid waste disposal, though plans and laws were quite different from state to state. In 1984, the Solid Waste Disposal Act was amended. The amendments added new regulations about disposing of waste. The regulations involved checking the effects of waste on groundwater, paying for cleanup of waste, inspections of waste sites, and other matters.

The Solid Waste Disposal Act was just a beginning. It was among the first of many government actions that were needed to control solid wastes, TOXIC WASTES, and HAZARDOUS WASTES. It was followed by such laws as the Resource Recovery Act of 1970, and the RESOURCE CONSERVATION AND RECOVERY ACT (RCRA). [*See also* COMPREHENSIVE ENVIRONMENTAL RESPONSE, COMPENSATION, AND LIABILITY ACT (CERCLA); HAZARDOUS SUBSTANCES ACT; INDUSTRIAL WASTE TREATMENT; RECYCLING, REDUCING, REUSING; TOXIC SUBSTANCES CONTROL ACT (1976); WASTE MANAGEMENT; and WASTE REDUCTION.]

Solid Waste Disposal Act

◗A United States law that was aimed at finding new ways to dispose of GARBAGE, refuse, and other discarded solid materials. In the 1960s, the problem of SOLID WASTE disposal was beginning to affect the entire United States. The amount of solid waste produced each year was increasing. This was partly because of changes in the way consumer products were made and

◆ Recycling materials is becoming more and more necessary as landfills fill up and as sites for new ones become scarce.

Solid Waste Incineration

▶ Burning of SOLID WASTES to reduce waste volume and, sometimes, to produce ELECTRICITY. Solid wastes consist of a mixture of materials including paper, PLASTICS, glass, and metal.

Burning solid wastes in incinerators can burn up the materials and reduce the volume of waste disposed of in LANDFILLS by 60%. Most incinerators are mass-burn incinerators that do not separate materials that will burn from those that will not (including hazardous materials). These types of incinerators have caused AIR POLLUTION problems because of the smoke and PARTICULATES they release into the ATMOSPHERE. In addition, new wastes, such as toxic FLY ASH, create problems related to disposal of the incinerator ash that may be hazardous.

Some incinerators, known as *waste-to-energy* plants, burn solid wastes to create steam and make electricity. There are about 150 waste-to-energy plants in the United States, but opposition to such new incinerators has been strong. Some states have banned solid waste incineration because of concerns about health and safety.

Effective solid waste incinerators have been built in several countries. The incinerators require careful attention in design, maintenance, and monitoring to assure that they help solve the solid waste problem rather than create new problems that are even more severe. [*See also* COGENERATION; GARBAGE; HAZARDOUS WASTE; LANDFILLS; SOLID WASTE DISPOSAL ACT; WASTE MANAGEMENT; and WASTE REDUCTION.]

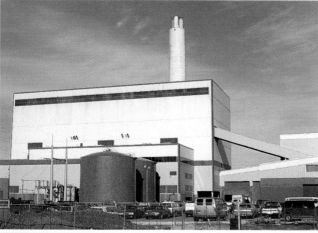

◆ The use of waste incinerators is banned in many cities. Appropriate pollution controls are expensive to install and maintain.

◆ The Vulcanus was built for ocean incineration of solid waste in order to avoid producing air pollution in cities.

Source Reduction

▶ Method of POLLUTION control that involves the reduction of the amount of SOLID WASTE produced by households and businesses. Solid waste is any discarded material that is not a liquid or gas.

About 87% of the solid waste produced in the United States comes from MINING and agriculture. Industries produce about 9% of the country's solid waste. The remaining 4%—paper, cardboard, PLASTIC, and glass—is used and thrown away by individuals or businesses. The waste produced by individuals and businesses is known as *municipal solid waste*. Source reduction involves limiting the use of a material that will have to be thrown away later, through recycling, reducing, and reusing.

To reduce the amounts of municipal waste dumped into LAND-FILLS, environmentalists support the idea of source reduction. Producing less waste saves money and energy by reducing the expense and difficulties involved in disposing of wastes.

Another way to produce less waste is by reducing the amount of material in a product. Reusing products more than once is another important method of source reduction. For example, the plastic containers and jars used to package foods can be used over and over for the storage of the leftover foods or for the storage of other items. [*See also* NONRENEWABLE RESOURCES;

◆ The recycling of newspapers not only reduces pressure on landfills but also saves trees.

◆ One example of source reduction is the use of a reusable net bag for groceries instead of disposable plastic bags.

RECYCLING, REDUCING, REUSING; RENEWABLE RESOURCES; RESOURCE CONSERVATION AND RECOVERY ACT (RCRA); SOLID WASTE DISPOSAL ACT; SOLID WASTE INCINERATION; and WASTE MANAGEMENT.]

Species

◗ A category of similar living organisms. In the case of those organisms that reproduce sexually, members of a species almost always cannot or do not usually produce fertile offspring by mating with other species. Members of a species have a number of common qualities, such as pattern or color, shape, size, body function, and behavior.

In some cases, although two organisms may seem similar to each other, there may be a slight difference in type of egg produced, behavior, or body chemistry that keeps them from belonging to the same species. This difference also means that they do not mate with each other. SUBSPECIES are organisms that have some different characteristics but still belong to the same species.

Species is the final grouping in the classification of living organisms, called *taxonomic groupings*. Scientists studying and sorting characteristics have divided living organisms into large groups called kingdoms. Kingdoms are divided into phyla or divisions; phyla into classes; classes into orders; orders into families; and families into genera (singular genus) and species.

TAXONOMIC GROUPS

Organisms are divided into five kingdoms based on their cell structure and the way they obtain food: monerans, protists, FUNGI, PLANTS, and animals. Monerans are single-celled BACTERIA and blue-green algae. Monerans have no nuclei in their cells. Some feed on other organisms; others can live in chemicals in the environment either by gaining energy from feeding or by using chemicals to produce energy, or through photosynthesis.

Protists are mostly single-celled organisms that have a nucleus in their cells. They feed by every known method, through photosynthesis or by being PREDATORS. Examples of protists are amoebas and paramecia. Fungi are plantlike organisms that feed directly on living or dead organisms with structures called *hyphae*.

The plant kingdom is broken down into divisions, classes, orders, and other classifications based on whether the plant reproduces by seeds, whether it makes cones or flowers, and the kind of flowers or seeds it makes. The animal kingdom classifications are based partly on whether the animal has a backbone and on such characteristics as how it reproduces and nurses its young, what it eats, and whether its body is covered with scales, feathers, or hair.

Within the animal kingdom, people are in the phylum chordata and the subphylum vertebrata because they have backbones. People are in the class mammalia because they have hair, are warm blooded, and have milk-producing glands for feeding their young. People are in the order primates because of the development of the skull, brain, teeth, and limbs. People are in the family Hominidae

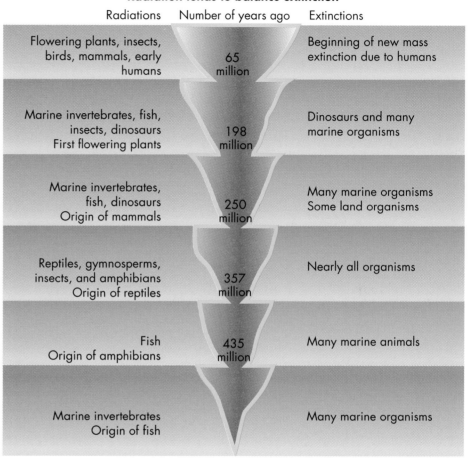

Radiation tends to balance extinction

Radiations	Number of years ago	Extinctions
Flowering plants, insects, birds, mammals, early humans	65 million	Beginning of new mass extinction due to humans
Marine invertebrates, fish, insects, dinosaurs First flowering plants	198 million	Dinosaurs and many marine organisms
Marine invertebrates, fish, dinosaurs Origin of mammals	250 million	Many marine organisms Some land organisms
Reptiles, gymnosperms, insects, and amphibians Origin of reptiles	357 million	Nearly all organisms
Fish Origin of amphibians	435 million	Many marine animals
Marine invertebrates Origin of fish		Many marine organisms

◆ Speciation tends to be balanced by extinction. The central part of the diagram shows when organisms arose or became extinct. The width of the central drawing relates to the total number of species living during the various time periods.

because of their larger brain, smaller jaws, and lower limb structure that enables them to walk on two legs. Finally, people are in the genus and species *Homo sapiens* because they have highly developed lower limbs for walking erect, skull development that includes a flat forehead, and a definite chin and neck.

SPECIES DIVERSITY

There are more than 1.5 million known species of organisms. Each year, 15,000 to 20,000 more species are discovered. Species of organisms that reproduce sexually—that is, a male and a female cell unite to form a new individual—are gradually changing. Each new individual has a new mix of inheritance (GENES)—one-half from each parent. This mix provides a new set of possibilities for development.

As reproduction continues and living conditions (the environment) change, the individual organisms that carry the best gene mix adapt to the environment and thrive. For

◆ There is more diversity in the insect class than in any other animal class.

example, a plant or animal that has the genes enabling it to live in a dry environment will thrive in that type of environment. It will pass on these useful qualities to its offspring. Those that do not have the genes needed to adapt die out. This is part of the process known as NATURAL SELECTION.

Some organisms produce new organisms asexually. A part of their body—perhaps a root or runner in the case of a plant—simply sprouts an individual just like the first. Some microorganisms, such as cer-

◆ These are different species in the plant family leguminosae.

tain bacteria, split in half to produce an exact copy of themselves, one with the same genes. Producing an exact copy means there is no new mixture in each individual, no new possibilities. Thus, organisms that reproduce asexually cannot adapt to the environment. For the most part they are changeless species. In rare cases, the genes of an asexual species may mutate, providing the species with the ability to change.

SPECIATION

Formation of a new species is called *speciation*. Speciation among sexual reproducers can occur when a group of individuals from one species becomes separated from other members of the species in a new geographical environment. This is called *allopatric speciation*. As the organism adapts to the environment and continues to breed, continually mixing and remixing possibilities in new individuals, it gradually changes. Soon it is so different that it cannot mate with members of the species to which it once belonged. It is a new species one more at the end of the classification that begins with this kingdom.

Genetic drift is another way in which a new species develops. This happens when a few members of a species carrying a gene combination rare to its species travel to another geographical area such as an island. There, the gene mix of their offspring will be quite different from that of most members of their species. In the new environment, natural selection will change this rare mix of genes to another species.

◆ The fish and the pocket mouse are adapted to different environments. The fish, adapted to living in a watery environment, has fins for swimming, scales for protecting it in the water, and gills to enable it to breathe underwater. The pocket mouse, a desert dweller, has fur to protect it from the sun, long hind legs for leaping through the sand, and lungs for breathing.

Sympatric speciation commonly occurs among INSECTS that live, feed, and mate on one kind of plant. Several insects move to another kind of plant growing nearby. Changes in genetic mix of the insects adapt them to life on the new plant. Soon the insect is so changed it cannot mate with members of its former species. It is a new species.

Polypoidy produces about half the new species of FLOWERING PLANTS. *Polypoidy* is the accidental doubling of the chromosome number of the plant during its early growth. It is an instant new species. The new polypoid chromosome number means the plant cannot breed with members of its former species that have only half as many chromosomes. Polypoidy also occurs in animals, though less commonly than it does in plants. [*See also* DARWIN, CHARLES ROBERT; EVOLUTION; GENETIC DIVERSITY; GENETICS; NATURAL SELECTION; and NICHE.]

Species Diversity

▮The number of different SPECIES of organisms within an area or BIOLOGICAL COMMUNITY. Species living together depend on each other for certain vital elements such as food and shelter. Thus, the greater the species diversity, the stronger the individual species and the more stable the community.

A community with many varieties is said to be species rich. Diversity is determined by the ENVIRONMENT. Species richness decreases as one moves away from the equator toward the poles. Species of BIRDS, for example, decrease from 1,400 in Colombia near the equator to 56 in Greenland near the North Pole. The same pattern is seen for MAMMALS, FISH, lizards, and trees. In the OCEAN, species richness increases as one moves from shallow water to deeper water.

There are more species in mountainous than in flat areas and fewer on peninsulas than on the adjoining mainland. There are more species of land birds and mammals in the western United States than in the eastern United States. However there are more species of toads, frogs, and salamanders in the moist, cool East, while there are more species of turtles, snakes, and lizards in the hot, dry West. Small islands have fewer species than large islands.

Why are there variations in species richness? It is clear that the environment affects species diversity. Scientists have a number of theories about how this happens. However, because they involve organisms in the wild, these theories are difficult to test.

Observations of environments with low species diversity provide some information. These environments include those where organisms are just becoming established and must endure such environmental stresses as severe hot or cold WEATHER; places where sudden changes in such elements as temperature or salt content can kill all life; and unpredictable environments where conditions change constantly. Such physical conditions as these, as well as the number of small environments within a larger area where life can take hold and the availability of food, are all factors that affect species diversity.

SPECIATION

Another important factor that plays a role in species diversity is *speciation,* the development of new species. New species often develop when a group of organisms becomes isolated from the larger group in a slightly different environment. An example is when a group of organisms are separated from other members of their species by a mountain or a stream. In such a case, the separated members can mate only with those in their isolated group. These individual organisms carry in their cells a mix of GENES that determine their form and how they function. When individuals mate, they each give one set of their genes to the offspring. The interaction of the two sets of genes from the parents may produce an offspring somewhat different from the parents. If this difference enables the offspring to cope with the environment better than other individuals, the offspring will survive and pass on those useful characteristics to its offspring. Others without these characteristics will die out.

This survival of the fittest, or NATURAL SELECTION, is the way in which one species evolves into another species. Thus, the group of organisms that found itself in a new environment has become a new species and increases species diversity in its area. Obviously the environmental change must be slight, because organisms do not possess the genetic capability to adapt to severe changes in the environment.

A look at some environments gives clues to elements that favor speciation. For example, in the TROPICS, food is abundant. Animals do not have to travel far for food. Groups of a species may become isolated and evolve into a new species. A mountainous area provides a wide variety of environments where species groups may find themselves isolated. Speciation is restricted in areas undergoing drastic changes in the environment such as the ocean surface.

Since each organism carries in its genes the possibilities for new organisms, EXTINCTION can destroy those possibilities for life, resulting in lower diversity. Extinction is caused by drastic environmental changes.

Physical changes in the environment such as severe temperature changes or increases in salt content can cause extinction. The total loss of any vital element—food, shelter, water—can also bring about extinction. These environmental changes can be natural, such as those that made dinosaurs extinct. Sometimes environmental changes can be caused by human activity, such as the destruction of a FOREST or the damming of a river. POLLUTION of food and water may poison the organism. Sometimes animals are hunted into extinction.

HUMAN IMPACT ON SPECIES DIVERSITY

Experts have given many reasons for protecting species diversity.

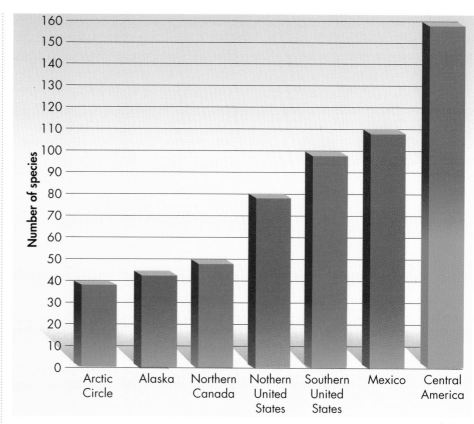

◆ Species diversity of mammals increases from cold polar areas to tropical Central America.

Many species are of direct or potential benefit to humans for food, medicines, and other products. A diversity of species is necessary for maintaining healthy and balanced ecosystems. Many people believe that it is important to protect species diversity from human impact as much as possible because they want their descendants to be able to enjoy other species, or because they believe that other species have a right to continued existence. [*See also* BIODIVERSITY; CARSON, RACHEL LOUISE; DARWIN, CHARLES ROBERT; ENDANGERED SPECIES; EVOLUTION; MASS EXTINCTION; WILDLIFE CONSERVATION; and WILSON, EDWARD OSBORNE.]

Spotted Owl
See NORTHERN SPOTTED OWL

Steppe
See PRAIRIE

Stratosphere

❿The middle layer of the ATMOSPHERE located above the TROPOSPHERE. The stratosphere extends

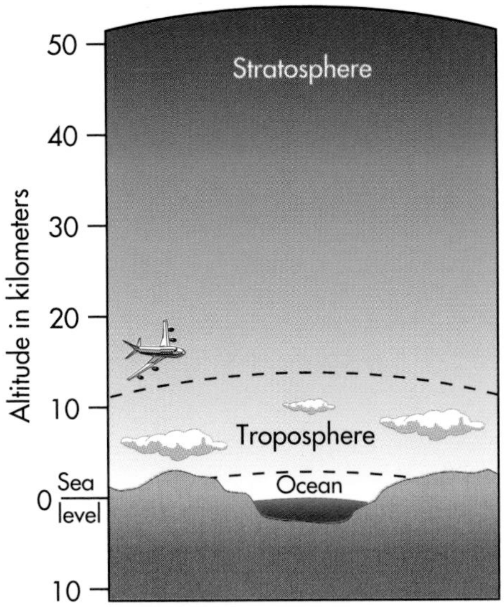

◆ The stratosphere contains Earth's ozone layer, a natural shield that protects life from the sun's dangerous rays.

from the top of the troposphere at about 6 to 10 miles (10 to 16 kilometers) from Earth's surface, to its upper limit at about 30 to 40 miles (50 to 55 kilometers) above Earth's surface. Temperatures rise slightly with altitude in the stratosphere.

The warming effect in the stratosphere is due to a layer of gas called OZONE. In the stratosphere, ULTRAVIOLET RADIATION from the sun split some two-atom OXYGEN molecules into single-atom oxygen molecules. These single-atom oxygen molecules combine with two-atom oxygen molecules to form ozone (O_3). Ozone is a three-atom form of oxygen.

Chemicals in the atmosphere can cause ozone to break down to normal oxygen. When it does, the stratosphere loses some of its ability to absorb ultraviolet radiation. This allows more ultraviolet radiation, which can be harmful to living things, to reach Earth's surface. The original ultraviolet light energy is given off as heat. The chemical action of the ozone layer heats the stratosphere. [*See also* AIR POLLUTION; CARCINOGEN; CLEAN AIR ACT; GLOBAL WARMING; GREENHOUSE EFFECT; MONTREAL PROTOCOL; OZONE HOLE; OZONE LAYER; and OZONE POLLUTION.]

Strip Mining

Method used for the extraction of COAL and MINERALS located on relatively flat terrain. Strip mining is a type of SURFACE MINING. In this type of MINING, bulldozers, power shovels, and trucks are used to remove SOIL and rock layers that lie above near-surface deposits of minerals or coal. The removed dirt, called *overburden,* is piled up alongside the exposed deposits. In the process, long, winding ridges and valleys that resemble a washboard are created on the land.

Strip mining is used to mine PHOSPHATES in Florida, North Carolina, and Idaho. However, the most common use of strip mining is for the mining of coal in the Appalachian Mountains, the Midwest, and the Southwest. Strip mining is usually done on flat terrain and is sometimes called *area mining.* However, strip mining in hilly areas, called *contour mining,* is also common.

Strip mining can cause serious damage to the ENVIRONMENT if done

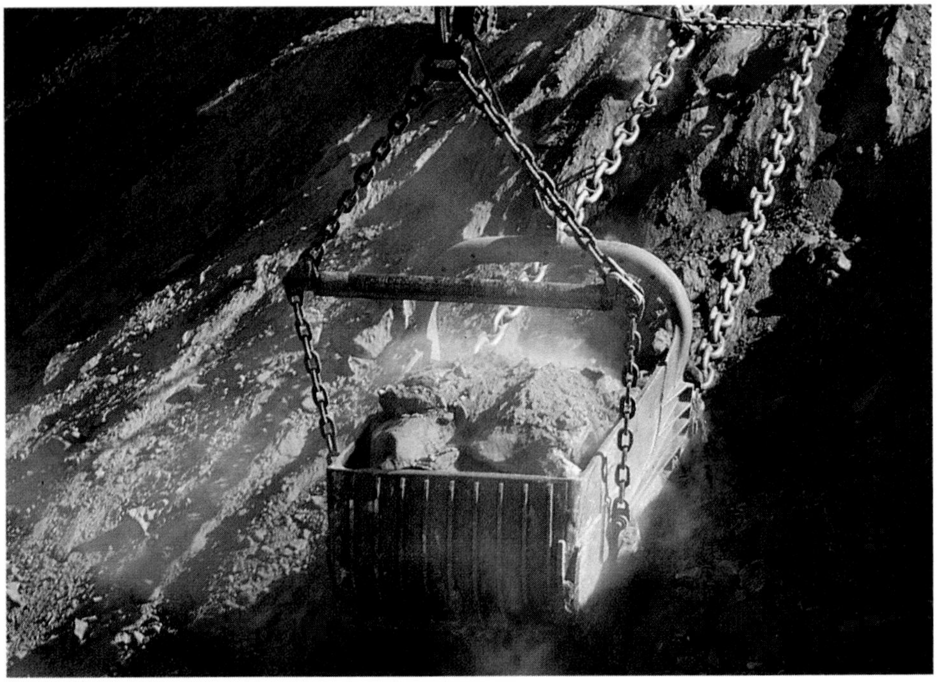

◆ A mining crane scoops up coal for copper smelting.

ment. The biggest problem seems to be an increase in soil EROSION that occurs when soils are loosened during the mining process. [*See also* HABITAT LOSS; HEAVY METALS POISONING; LAND USE; MINERAL LEASING ACT; MINING LAW OF 1872; RECLAMATION ACT OF 1902; and TAILINGS.]

Subsistence Agriculture

▌Farming practice in which crops and LIVESTOCK are raised to provide enough food for a family. Subsistence farms are typically small, consisting of only a few acres and simple equipment. There is little, if any, food left for trade and surplus.

Subsistence farming is a traditional farming practice that has been used throughout the world ever since agriculture was first developed in the Middle East about 10,000 to 11,000 years ago. Early subsistence farmers did not use PESTICIDES or any sophisticated machinery to raise their crops. Instead, they often moved from plot to plot as soon as the SOIL became depleted of nutrients.

Subsistence farming persists today in many parts of the world, particularly in underdeveloped

carelessly. The most immediate disturbances involve the disruption of ECOSYSTEMS. Strip mining removes all vegetation from an area, which destroys the HABITATS of countless organisms. Strip mining also contributes to WATER POLLUTION when RUNOFF from mined areas introduces acidic water and toxic materials into groundwater supplies.

In 1977, Congress passed the SURFACE MINING CONTROL AND RECOVERY ACT (SMCRA). This law requires mining companies to restore mined lands as near as pos-

sible to their original conditions. This process, known as reclamation, involves filling the mined area with TOPSOIL and replanting the area with grasses and trees. Mined areas also are monitored for many years to make sure the ecosystem is regenerating properly.

If done correctly, reclamation of a mined area can return an ecosystem close to its original condition. However, a study by the U.S. Geological Survey in 1970 concluded that strip mining can have long-term consequences for the environ-

◆ Much of the labor required by agriculture in poorer countries is performed by animals.

◆ The percentage of the population employed in agriculture is higher in countries where there is subsistence agriculture than where agricultural processes are mechanized.

regions, such as eastern and western Africa, northern Mexico, eastern Brazil, and many parts of Asia. The methods of subsistence agriculture have not changed much over the course of history. Many subsistence farmers use ORGANIC FARMING techniques such as CROP ROTATION to increase soil fertility and prevent EROSION, BIOLOGICAL CONTROL to eliminate INSECTS, and other natural methods to increase food production while limiting costs. [*See also* AGRICULTURAL REVOLUTION; AGROECOLOGY; GREEN REVOLUTION; IRRIGATION; NO-TILL AGRICULTURE; and TERRACING.]

Subspecies

Populations of a SPECIES that differ slightly in appearance and behavior from each other. Just as individuals within a population differ from one another, variation also exists among populations. When populations in different geographical areas show slightly different characteristics, scientists often classify them as subspecies.

In the United States, an example of subspecies classification is seen in rat snakes (*Elaphe obsoleta*). Rat snakes are small snakes that live throughout the eastern and central United States. There is a great variety in the body color (brown, yellow, orange, or black) and stripe pattern of rat snakes that seems to correspond to where they live. Scientists believe the differences in

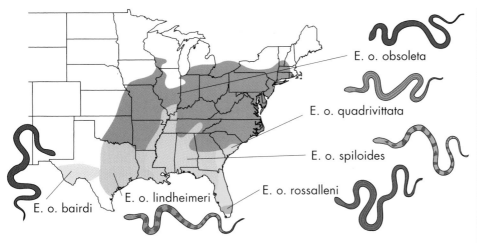

The various types of rat snakes in the United States are classic examples of subspecies. Scientists believe that the differences are due to the environment.

E. o. obsoleta

E. o. quadrivittata

E. o. spiloides

E. o. rossalleni

E. o. bairdi

E. o. lindheimeri

rat snakes are related to differences in the ENVIRONMENT. Some subspecies are threatened or endangered while other species may not be. The ENDANGERED SPECIES ACT allows rare subspecies to be legally protected. [*See also* ADAPTATION; BIODIVERSITY; EVOLUTION; NATURAL SELECTION; NORTHERN SPOTTED OWL; and SPECIES DIVERSITY.]

Succession

❙A natural sequence of SPECIES replacement in an ECOSYSTEM, such as that which occurs as a meadow gradually becomes a FOREST. Succession is often difficult to observe. It can take decades, sometimes centuries, for one BIOLOGICAL COMMUNITY to completely succeed another. Eventually, a final, stable community, called a CLIMAX COMMUNITY, may form if the area is left undis-

turbed. Climax communities continue to change in small ways, but they remain generally similar over long periods of time. For instance, a

mature beech-maple forest will remain a beech-maple forest as long as no serious disruptions occur within the community.

Why does succession occur? Over time, the species in a biological community make conditions suitable for organisms that make up another community. When the new species move in, they out-compete the previous species for resources such as space, water, and sunlight. The new species eventually replace the previous ones.

TWO TYPES OF SUCCESSION

Ecologists recognize two types of succession. *Primary succession* refers to succession in previously barren areas, such as on the rocks of a newly created island, a beach,

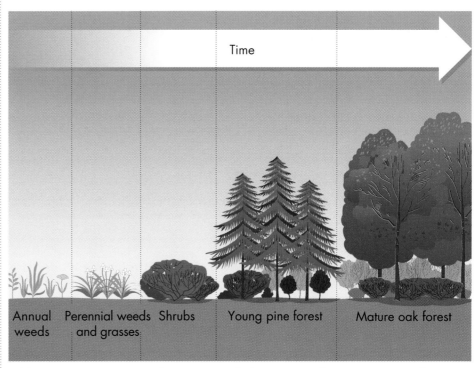

Annual weeds

Perennial weeds and grasses

Shrubs

Young pine forest

Mature oak forest

◆ New ecosystems arise either in previously barren areas (primary succession) or in areas destroyed by natural disasters (secondary succession).

◆ The catastrophic eruption of Mount St. Helens wiped out acres of plants. Secondary succession will occur on the site, restoring the ecosystem.

rapidly reproducing species, such as mosses, LICHENS, and BACTERIA. They are able to withstand harsh environmental conditions.

Consider the changes that occur on new rocks formed by volcanic eruptions. The first pioneer species to colonize or move into the bare rock will probably be bacteria and lichens, organisms that can live without SOIL. Lichens, the crusty, multicolored organisms that often grow on rocks and tree trunks, are important pioneers because they release chemicals that can slowly break up rock to form SOIL. Water may also freeze and thaw in tiny cracks, further breaking up the rock. Mosses may move in and take hold on the rocks, breaking up the rocks even more. As lichens, mosses, and bacteria die, their decaying bodies accumulate in the cracks and crevices, aiding in the process of soil formation.

When there is a sufficient amount of fertile soil, organisms such as FERNS, weeds, and INSECTS move in. These species eventually replace the pioneers by out-competing them for key resources. As all these organisms die and decay, they add to the growing pile of soil. Seeds borne by the wind blow into these larger patches of soil and begin to grow. Over time, new HABITATS emerge and new species move in. Eventually, the area may become a climax community containing mature trees, shrubs, and a variety of forest-dwelling animals.

or the concrete surfaces of an old playground or sidewalk. *Secondary succession* refers to succession in areas that have undergone primary succession or areas disrupted by NATURAL DISASTERS or human activities. In both types of succession, species enter and leave the bio-

logical community in a regular sequence.

Primary Succession

Primary succession begins when PIONEER SPECIES move into a barren area. Pioneer species are hardy,

Secondary Succession

Secondary succession occurs when the dominant PLANT species of a community, often mature trees, are removed. Natural disasters, such as FOREST FIRES and volcanic eruptions, can quickly remove the trees from an area. Human activities, such as logging and CLEAR-CUTTING, also remove trees.

Secondary succession happens much more rapidly than primary succession, because fertile soil is usually present. Seeds, worms, insects, and other organisms may also be present. In other words, the new biological community is not "starting from scratch."

When forest fires destroyed nearly 40% of YELLOWSTONE NATIONAL PARK during the summer of 1988, ecologists observed secondary succession almost immediately. The first plants to grow in the bare soil were wildflowers and weeds, plants that reproduce quickly and grow well in open sunlight. Within three years, other plants—grasses, ferns, and pine seedlings—were seen sprouting up through the wildflowers. Today, the pine seedlings continue to grow. Eventually, these seedlings will block out enough light so that grasses will no longer grow. A mature forest will once again develop.

The process of secondary succession is an important part of forest management. Foresters sometimes let natural fires burn because fires are important to the maintenance of many forests. For example, some plant species, such as the jack pine, grow some cones that release seeds only after they have been exposed to the heat of a fire. [*See also* ABIOTIC FACTORS; COMPETITION; DECOMPOSERS; DECOMPOSITION; FIRE ECOLOGY; FORESTRY; PRESCRIBED BURN; VOLCANISM; and WEATHERING.]

Sulfur Dioxide

▶An air pollutant produced mainly by burning FOSSIL FUEL with a high sulfur content. Sulfur is an element that is present in organisms, so it is also present in the type of fuel made of incompletely decomposed organisms. It is very abundant in COAL; 70% of the sulfur in the air is emitted by electricity-generating power plants that use coal. Oil contains some sulfur, and NATURAL GAS contains little sulfur.

EFFECTS OF POLLUTION

When sulfur is released from fuel by burning, it combines with OXYGEN in the air to produce sulfur dioxide (SO_2). When sulfur dioxide reacts with water it forms sulfuric acid. This acid harms organisms and damages nonliving things like stone and rubber. The acid tends to destroy tissues of the animals' breathing tracts, especially when there is soot in the air, because the sulfur compound combines with the soot and is brought deep into the lungs. Sulfuric acid is also very corrosive and eats away at the surfaces of statues, destroying them in the process.

ACID RAIN

Both sulfur dioxide and NITROGEN DIOXIDE can combine with water vapor in the ATMOSPHERE to form acids. These acids fall to Earth as ACID RAIN. Much acid rain falls into lakes and streams, making the water in these bodies acidic. When water turns acid it can kill the organisms that live in it, either because they are sensitive to acidity or because the acidic water leaches toxic levels of metals into the water. Despite efforts to control the production of sulfur dioxide, acid rain POLLUTION in Eastern Europe, China, Russia, India, and many developing countries is worsening.

REDUCTION OF SULFUR POLLUTION

One method of reducing the amount of sulfur dioxide released by burning FUELS is to remove the sulfur before the fuel is burned or to make use of fuel that contains little sulfur. Fuels contain different amounts of sulfur. There is high-sulfur fuel and low-sulfur fuel. However, low-sulfur fuel is very expensive. Removing the sulfur from high-sulfur fuel is also expensive.

Another method used to reduce sulfur emissions is to place pollution-control devices in smokestacks. One such device is called a SCRUBBER. A drawback to the use of pollution-control devices in smokestacks is that they have to be cleaned and maintained. Thus, they add to the expense of fuel use. [*See also* AIR POLLUTION; CATALYTIC CONVERTER; CLEAN AIR ACT; HYDROCARBONS; NITROGEN OXIDES; and WEATHERING.]

Superfund

▶The COMPREHENSIVE ENVIRONMENTAL RESPONSE, COMPENSATION, AND LIABILITY ACT (CERCLA), signed into law by

President Carter in 1980. CERCLA is often referred to as the Superfund. The Superfund empowered the ENVIRONMENTAL PROTECTION AGENCY (EPA) and other federal agencies to make emergency cleanups of areas where there was a release, or would likely be a release, of HAZARDOUS WASTES. Hazardous waste pose a threat to human health and the ENVIRONMENT. The cost of each cleanup, according to CERCLA, would be paid jointly by the EPA, the state in which the site was located, and the company or persons responsible for the POLLUTION problem.

The money for Superfund's original budget of $1.6 billion came from an additional tax placed on crude oil and selected chemicals. The budgeted funds allowed the EPA to pay for any cleanup activities and then attempt to collect the money from responsible parties. The Superfund Amendments and Reauthorization Act of 1986 (SARA) increased the budget to $8.5 billion. It also strengthened EPA's ability to enforce policies emphasizing TOXIC WASTE reduction, to list all toxic emissions, to get more cleanup accountability at government facilities, and to develop local-level emergency plans for handling a potential hazardous waste crisis.

The Toxic Release Inventory (TRI), established by SARA, monitors the release of toxic emissions by businesses with more than ten employees. In 1989, TRI reported that such businesses had released 2.4 billion pounds (1 billion kilograms) of toxic chemicals into the air and 189 million pounds (86 million kilograms) into streams, lakes, and rivers.

◆ EPA workers use bulldozers to clean up a Superfund site in Orange County, California.

LISTING CLEANUP SITES

The EPA set up a National Priorities List (NPL), ranking U.S. sites eligible for cleanup under Superfund's federally allocated money. A site was ranked according to its possible harm to humans and the environment from the movement of hazardous materials by water or air; by explosion and fire; or from direct contact at the disposal facility. Other important factors include the kind and amount of waste, the amount of pollution already present—especially in groundwater—and the number of people living and working near the site.

In 1980, about 8,000 sites were studied and listed. Since then, many more have been added. Every state has at least one site on the NPL.

Some have many more than one. For example, New Jersey has about 110, Pennsylvania 97, and New York, Michigan, and California each has more than 75 hazardous waste sites on the EPA's list. As of February 1991, EPA's NPL had a total of 1,189 sites.

The General Accounting Office (GAO) reports that there are between 130,000 and 425,000 hazardous waste sites in the United States that must be reviewed. The EPA does not agree; it says the number is closer to 30,000.

FROM PROMISE TO PRIORITY

Superfund came into being as a result of public outcry in the late

1970s about hidden threats from toxic waste dumps. Several tragedies in the news gave the public living examples of the danger.

A LANDFILL in the LOVE CANAL area of Niagara Falls, New York, had been used as a dump site for chemicals in the 1940s and 1950s. When a housing development was built there years later, smelly chemicals oozed from the ground. Residents began suffering from higher than normal occurrences of miscarriages, birth defects, and CANCER. A 1978 investigation found that the area's groundwater had been contaminated by toxic waste from the landfill. The residents of Love Canal were moved out of the town, which was declared unfit for human existence.

In 1980, a New Jersey warehouse used as a hazardous waste dump exploded and exposed hundreds of thousands of people to contamination. These and other disasters prompted Congress to pass the law that created Superfund. The law ordered the EPA to clean up such abandoned dumps, including 24 that were said to pose a greater danger to the public health than Love Canal.

Many people believed Superfund held a promise for a cleaner, safer environment. However, the promise has not been fulfilled. During the early part of President Reagan's administration (1980–1988), the Superfund program was at the center of a political scandal. Congress was investigating allegations that the EPA misused Superfund money. The EPA was also accused of making "sweetheart deals" with some polluting businesses, allowing them to continue dumping toxic waste in areas needing cleanup. Congress requested official EPA Superfund documents, but even the president would not release them. Most of the important documents were shredded. Congress and citizens became distrustful of Superfund executives, some of whom were fired. Although newly hired executives promised changes, progress has been slow.

The Office of Technology Assessment (OTA) has expressed concern that Superfund is not working the way Congress intended. There have been relatively few cleanups, and those few often used inadequate methods, leading to a lack of public trust. By the end of 1988, some 433 sites had been recognized for cleanup and were awaiting action. Corrective work was being done at 177 sites, but final cleanup had been achieved at only 32 out of a total of 1,189 sites listed.

The Superfund program also had problems hiring, training, and keeping capable staff members. Cleanup costs were estimated to be about $10 million per site. The total cost for the cleanup of all hazardous waste sites may reach $500 billion over the next fifty years. That figure does not count federal costs of cleaning up waste sites belonging to the Department of Energy.

PROS AND CONS

Superfund is a controversial part of the U.S. environmental plan. It has not cleaned up many sites in the more than 15 years it has been in action. In addition, getting polluters to pay the bill for their past contamination of land, air, and water has been difficult.

On the plus side, Superfund has helped to stop the procedures that originally produced toxic waste

◆ Illegally dumped toxic waste is removed from a backyard by EPA workers.

sites. Superfund regulations made landowners responsible for damages when it determined that owners were responsible for hazardous waste on their land, even if they did not dump it there themselves. Today, major land buyers regularly check their properties for signs of dumping or spills. Many banks now demand careful investigations of any land, looking for signs of hazardous materials, before making loans to consumers wishing to buy land or invest in land developments. [*See also* DIOXIN; ENVIRONMENTAL ETHICS; HEALTH AND DISEASE; INDUSTRIAL WASTE TREATMENT; MEDICAL WASTE; PCBS; RADIOACTIVE WASTE; TOXIC WASTE, INTERNATIONAL TRADE IN; WASTE MANAGEMENT; and WASTE REDUCTION.]

Surface Mining

▮The process of extracting COAL and MINERALS that lie close to Earth's surface. There are two general ways of removing substances from Earth: surface mining and subsurface mining. Subsurface mining is used when a substance is well below Earth's surface. Long, deep shafts are dug using explosives that blast out the rock.

In surface mining, bulldozers, power shovels, and trucks remove layers of SOIL and rock that lie over a vein of coal or a deposit of another mineral. The dirt, or *overburden*, is then piled up on one side of the deposit and the substance is extracted. After the MINING

operation is complete, the land can be restored.

Surface mining is a popular mining technique because it is less costly than subsurface mining. It is also much more efficient than subsurface mining because most of the deposit can be removed. In the United States, a large amount of coal is removed by surface mining. This method is most economical when the deposits lie close to the surface. However, surface mining equipment has the capability of removing overburden up to 200 feet (60 meters) thick.

OPEN-PIT MINING is a surface mining technique used on relatively flat terrain. In this method, gigantic power shovels dig long, winding parallel trenches to expose mineral and coal deposits. In some ways, the process is similar to a farmer plowing furrows in a field. Open-

pit mines can reach fantastic sizes. One open-pit COPPER mine in Utah, for instance, is nearly 2 miles (3.2 kilometers) long.

Surface mining can cause serious environmental damage if done in a careless manner. The most obvious disturbance is the total clearing of trees, shrubs, flowers, and small PLANTS in an area. When vegetation is cleared, the HABITATS of many organisms are destroyed and the ECOSYSTEM is disrupted. Surface mining can also pollute water sources when acid water RUNOFF from the mining area contaminates underground water supplies.

Federal laws require that TOPSOIL must be replaced in a mined area after mineral extraction. However, many people do not think that this requirement is enough because replaced soils are usually dry and infertile. In addition, the

◆ Surface mining of coal causes serious environmental damage to ecosystems.

salts, heavy metals, and acids exposed during mining generally make the area unsuitable for plant growth for long periods of time. [*See also* HABITAT LOSS; HEAVY METALS POISONING; LAND USE; MINERAL LEASING ACT; MINING LAW OF 1872; OFFICE OF SURFACE MINING, RECLAMATION, AND ENFORCEMENT; RECLAMATION ACT OF 1902; and STRIP MINING.]

Surface Mining Control and Recovery Act (SMCRA)

❿Federal law that requires companies to restore the ENVIRONMENT after removing MINERALS mined at the surface. SURFACE MINING is a method used to remove COAL, COPPER, and other minerals from underground deposits that lie near Earth's surface.

In surface mining, bulldozers and power shovels remove layers of dirt and rock that cover near-surface deposits. This practice involves large-scale removal of vegetation from an area and can damage the environment by permitting soil EROSION to occur rapidly. In 1977, Congress passed the Surface Mining Control and Recovery Act, a law that requires companies to restore or reclaim mined lands to their near-original conditions.

Most surface mining operations in the United States involve the

◆ The Surface Mining Control and Recovery Act requires mining companies to restore public lands to near-original conditions following mining activities.

removal of coal. Before 1977, millions of acres of disturbed land in the United States had been abandoned by coal companies and not restored. The 1977 law made the following four important points to address the environmental problems associated with surface mining:

1. Surface-mined lands must be returned to approximately the same conditions they were in before MINING.

2. Surface mining can be done only on certain PUBLIC LANDS.

3. Mining companies must use the best technology available to reduce POLLUTION of water supplies due to RUNOFF from mining fields.

4. Individual states are responsible for enforcing the law.

Since 1977, the Surface Mining Control and Recovery Act has been successful at protecting ECOSYSTEMS that have been disrupted by mining activities. However, in recent years, amendments and a lack of funding for federal inspection teams have reduced the effectiveness of this law. [*See also* HABITAT LOSS; HEAVY METALS POISONING; LAND USE; MINERAL LEASING ACT; OFFICE OF SURFACE MINING, RECLAMATION, AND ENFORCEMENT; and RECLAMATION ACT OF 1902.]

Surface Water

⏵Fresh water located on Earth's surface. The world's freshwater supplies can be divided into two categories: groundwater and surface water. Water that seeps into SOIL is called *groundwater*. Most groundwater collects in AQUIFERS. Less than 1% of water on Earth is groundwater. However, groundwater supplies most of the world's fresh drinking water. In contrast, surface water is the fresh water contained in lakes, ponds, rivers, streams, and puddles.

Surface water is in a more limited supply than groundwater. It is used as a source of drinking water in many parts of the world. Surface water also provides HABITAT to a variety of PLANTS, animals, and microorganisms and provides humans with areas for recreation and transportation. In addition to its use by organisms, surface water plays a key role in the global WATER CYCLE. For these reasons, environmentalists have sought to protect surface waters from the harmful effects of POLLUTION and development.

THREATS TO SURFACE WATERS

Throughout history, humans have settled where surface water was abundant. Because of this, surface waters, as well as OCEANS, have always faced the threat of pollution resulting from human activity. Direct pollution from a single, or POINT SOURCE, such as a factory, a SEWAGE TREATMENT PLANT, or a boat, is one of the greatest dangers to the organisms that live in the world's surface waters. Industrial WASTEWATER, for instance, may contain a variety of pollutants, including heavy metals, acids, oil, organic materials, and heat. When discharged into a river or lake, the wastewater can poison organisms and disrupt normally healthy ECOSYSTEMS.

An even greater threat to surface waters is nonpoint pollution. Also known as *pointless pollution,* nonpoint pollution comes from many sources rather than a single specific site. Much nonpoint pollution reaches surface waters when rain washes PESTICIDES, gasoline, oil, household chemicals, and other toxic substances into sewer systems or water courses. This polluted water eventually reaches lakes and rivers where it can damage ecosystems. Because nonpoint pollution enters bodies of water in many different ways, it is very difficult to control, regulate, and prevent.

PROTECTING SURFACE WATERS

Because of the importance of surface waters to humans and other SPECIES, most industrialized nations have developed laws to protect these waters. In the United States, the CLEAN WATER ACT is the main piece of legislation aimed at protecting surface waters. Passed by Congress in 1972, the stated purpose of the Clean Water Act is to "restore and maintain the chemical, physical, and biological integrity of the nation's waters." Its goal is to make all surface waters clean enough for swimming, FISHING, and drinking.

Since 1972, many states have passed their own water quality laws. Together, federal and state laws have had many positive effects on the quality of surface waters. For example, in many areas toxic chemicals are now removed from wastewater. Many industrial wastes must be treated and cleaned prior to disposal. However, environmentalists argue that water quality laws need to be even more strict. Individuals and businesses need to cooperate to reduce the amount of nonpoint pollution that enters the world's surface waters. [*See also* ACID RAIN; ALGAL BLOOM; BIOACCUMULATION;

CHLORINATION; EFFLUENT; EUTROPHICATION; GREAT LAKES; INDUSTRIAL WASTE TREATMENT; NONPOINT SOURCE; RUNOFF; SAFE DRINKING WATER ACT; SEDIMENTATION; THERMAL WATER POLLUTION; WATER, DRINKING; WATER POLLUTION; WATER QUALITY STANDARDS; and WATER TREATMENT.]

Survival of the Fittest

See NATURAL SELECTION

Sustainable Agriculture

▶Farming practices that help protect the ENVIRONMENT by limiting the use of water, energy, synthetic PESTICIDES, and chemical fertilizers. Sustainable agriculture is similar to ORGANIC FARMING, but on a larger scale.

Sustainable agriculture practices use the latest advances in technology combined with traditional farming methods to limit damage to the environment. For instance, CROP ROTATION, the practice of alternating different crops on a plot of land, helps maintain soil fertility when crops that add nutrients to the SOILS are planted.

INTEGRATED PEST MANAGEMENT (IPM) is another important part of sustainable agriculture. IPM is a system of pest management that uses a combination of methods to help reduce pest problems. For instance,

instead of using only synthetic PESTICIDES to control pests, farmers practicing sustainable agriculture would try to use the natural enemies of the pests to control them. This practice is known as using BIOLOGICAL CONTROL.

Sustainable agriculture tries to establish a long-term relationship with the environment by limiting the harmful effects of other farming methods. For example, farmers using sustainable agriculture protect the environment by not making extensive use of chemical fertilizers that might damage the environment. Similarly, they make use of farming methods that help prevent EROSION of TOPSOIL. Today, the U.S. DEPARTMENT OF AGRICULTURE offers funding to farmers and researchers to help them develop new sustainable agriculture methods. [*See also* AGROECOLOGY; BIOACCUMULATION; DDT; NO-TILL AGRICULTURE; PEST CONTROL; and SOIL CONSERVATION.]

Sustainable Development

▶Goals for living in such a way as to ensure long-term stability of the ENVIRONMENT and CONSERVATION of NATURAL RESOURCES. Sustainable development is not a single environmental policy. Rather, it is a list of priorities and goals set by most industrialized nations today. It is a dynamic process designed to meet today's needs without harming the

ability of future generations to meet their needs.

Most people recognize that Earth's natural resources, such as oil, COAL, MINERALS, clean water and air, and even PLANTS and animals, are either in limited supply or might be damaged by pollution even if the resource is a RENEWABLE RESOURCE. People are also concerned about the environmental damage caused by the overuse of particular resources, such as FOSSIL FUELS.

Most scientists and experts agree that if humans as a SPECIES are to survive and prosper in the future, we must change the way we interact with the environment. Such changes include increasing the world's food supply and developing new techniques that can produce quality crops while limiting damage to the environment; improving and developing cleaner and more efficient FUELS; slowing POPULATION GROWTH; developing renewable energy sources such as SOLAR ENERGY, WIND POWER, and HYDROELECTRIC POWER; limiting the release of GREENHOUSE GASES and other pollutants by improving clean air technology; reducing waste generation by recycling; and protecting OCEANS and coastal resources.

Another major goal of sustainable development is to protect Earth's plants and animals and curb the loss of SPECIES. Many nations around the world have strict environmental policies aimed at conserving BIODIVERSITY. In the United States, for instance, the ENDANGERED SPECIES ACT is the main body of law that protects threatened and ENDANGERED SPECIES from human activities. Since its passage in 1973, the num-

Principles of Ecosystems Sustainability	
For Sustainability	• Ecosystems dispose of wastes and replenish nutrients by recycling all elements.
	• Ecosystems use sunlight as their source of energy.
	• The size of consumer populations is maintained such that overgrazing and other forms of overuse do not occur.
	• Biodiversity is maintained.

ber of species protected by this act has grown to over 900. More than 4,000 species are also under review and being considered for endangered or threatened status. Wolves, manatees, Florida panthers, BALD EAGLES, and whooping cranes are just a few of the many animals that have benefited from the Endangered Species Act. [*See also* AGROECOLOGY; AGROFORESTRY; ALTERNATIVE ENERGY SOURCES; BIOLOGICAL CONTROL; CROP ROTATION; ENVIRONMENTAL EDUCATION; ENVIRONMENTAL ETHICS; FRONTIER ETHIC; INTEGRATED PEST MANAGEMENT (IPM); NONRENEWABLE RESOURCES; ORGANIC FARMING; RECYCLING, REDUCING, REUSING; RENEWABLE RESOURCES; and SUSTAINABLE AGRICULTURE.]

Goals for Sustainable Development		
Some Ends	Elimination of waste and pollution	
	Preservation of Earth's life-sustaining processes for all people, now and in the future	
	World peace	
Some Means	1.	End government subsidy of wasteful, polluting economic activities.
	2.	Use government subsidy to encourage economic activities that conserve resources and produce less pollution.
	3.	Encourage recycling and reuse of materials.
	4.	Do more with less.
	5.	Exploit renewable energy sources at sustainable rates.
	6.	End deforestation, overgrazing, groundwater depletion, and soil erosion.
	7.	Educate people to understand our planet's life-sustaining processes.
	8.	Control human population growth.
	9.	Encourage global economic and political cooperation.
	10.	Distribute the world's wealth and resources more fairly.

Symbiosis

▶The close association between two organisms of different SPECIES that benefits at least one of the organisms. There are three main types of symbiotic relationships. They are MUTUALISM, COMMENSALISM, and PARASITISM.

MUTUALISM

Mutualism is a symbiotic relationship in which both organisms benefit and often cannot exist without each other. LICHENS provide an example of a mutualistic partnership. Lichens are an organism composed of a fungus and an alga. The ALGAE are protists that carry out PHOTOSYNTHESIS to produce food for the FUNGI. The fungi provide structure, support, and protection for the algae. Neither species could survive without the other.

Oxpecker BIRDS and rhinoceroses have a different kind of mutual relationship. In this relationship, the oxpeckers feed on INSECT pests attracted by rhinoceroses. The birds also warn the rhinoceroses of potential danger. The birds benefit from the relationship by being provided with a home and a fairly consistent food supply.

◆ Mistletoe is a parasite of various trees species.

◆ Oxpeckers and giraffes have a mutualistic relationship. The birds eat insects that bother giraffes. Thus, giraffes indirectly provide the oxpeckers with food; in return, the giraffes are relieved of their pests.

♦ Spanish moss has a commensal relationship with trees. It receives support and access to light, and the trees are neither helped nor harmed.

COMMENSALISM

Commensalism is a symbiotic relationship in which an organism benefits while the other remains relatively unaffected—being neither helped nor harmed. Seeds with spines or barbs form short-term commensal relationships with humans and other animals that aid in their dispersal. Sticking to fur or clothing, these seeds are often carried long distances without being more than mildly annoying to their hosts.

Spanish moss, which grows on tree branches in the southern part of the United States, forms another type of commensal association with its host. In this commensal relationship, the Spanish moss receives more support and has more access to light than it would if it grew on the ground. The tree on which Spanish moss lives is neither helped nor harmed by the relationship.

PARASITISM

Parasitism is a symbiotic relationship in which one organism benefits while the other is harmed in some way. Parasites are involved in a predator-prey relationship with their hosts. The parasitic organism, which depends on its host for nutrition and other needs, may sometimes kill its host. However, such an occurrence is rare because killing the host would eliminate the parasite's food supply. Tapeworms that live in the digestive tracts of humans and other animals are parasitic organisms. These parasites feed on materials that the host organism uses for food. The mistletoe that grows on trees is also a parasite. In this parasitic relationship, the mistletoe seedlings sink their roots into tree branches, taking nutrients from the sap of the tree.

IMPORTANCE OF SYMBIOSIS

Symbiotic relationships support the life activities of many organisms. More than 90% of all land plants rely on an association between their roots and fungi that make soil nutrients available to the plant. One of the most far-reaching examples of symbiosis is the association between leguminous plants, such as alfalfa and clover, and NITROGEN-FIXING root BACTERIA. This relationship plays a major role in the NITROGEN CYCLE.

Plant-eating insects, such as termites, and MAMMALS could not break down the **cellulose** in plant cell walls or absorb nutrients without the aid of microorganisms living in their digestive tracts. Bacteria, called *E. coli*, that live in the digestive tracts of humans and animals prevent the excessive growth of yeast and microorganisms that would otherwise cause intestinal upset. These bacteria also aid in the production of certain essential vitamins, such as biotin. [*See also* VIRUS.]

Synthetic Fuel

▶**A**rtificially-made liquid or gaseous FUEL produced from COAL. Synthetic fuels, or *synfuels,* have several advantages over coal. They are less costly and easier to transport, and are much less damaging to the ENVIRONMENT when burned for energy. In fact, synfuels can usually be burned without expensive pollution-control devices. However, because production of synthetic fuels is very expensive, there is currently not a high demand for them. Synfuels can also be made from OIL SHALE and BIOMASS (animal and plant wastes).

MAKING SYNFUELS

The production and use of coal-based synthetic fuels emerged in Germany during World War II. Germany faced a blockade of oil supplies and used new technology to make gasoline from coal. Liquid and gaseous synfuels are made when hydrogen is added to coal at a very high temperature and pressure to produce HYDROCARBONS.

Coal gasification is the process of converting coal to gas that can be burned more cleanly as a fuel than the coal itself. Unlike coal, the synfuel can be transported cheaply through a pipeline.

In one method of coal gasification, coal is first converted into coke, a nearly pure form of CARBON. The coke is then reacted with OXYGEN to form a gaseous mixture of hydrogen and CARBON MONOXIDE. This highly combustible gas, known as *producer gas,* does not produce much heat when burned, and thus is not very economical to transport long distances. Instead, it is usually used at the production site for heating and to generate ELECTRICITY.

A second, more useful type of gasification process converts coal

to synthetic natural gas, or SNG. SNG is chemically similar to NATURAL GAS and has a high heating value. Therefore, it is very economical to transport by pipeline. SNG is produced when powdered coal is burned in the presence of air. The products of this reaction are then combined with hydrogen gas to form METHANE, the chief compound of natural gas.

Coal liquefaction, or the process of turning coal into liquid, involves converting coal into liquid hydrocarbons such as methanol or synthetic gasoline. Introduced in Germany, this process is now successfully used in South Africa to produce a large portion of its liquid fuel needs.

ADVANTAGES AND DISADVANTAGES OF SYNFUEL

Although synthetic coal-based fuels are cleaner and cheaper to transport than solid coal, there are some drawbacks to large-scale use of synfuels. First, it is much more expensive to build a synthetic fuel plant than it is to build a coal-burning plant with state-of-the-art pollution-control devices. Another problem is that synfuels contain only 30%–40% of the energy content of solid coal; thus large-scale production would rapidly deplete world coal supplies.

During the 1970s, the United States was much more committed to the use of synfuels than it is

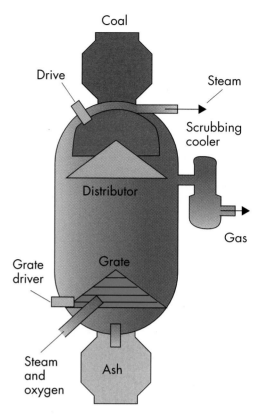

◆ Coal gasification, the transformation of coal into gas, requires energy.

now. In 1980, Congress created the U.S. Synthetic Fuels Corporation to manage and promote the development of synfuels technology. Eighty-eight billion dollars in grants was offered to oil companies to develop synfuels. However, because companies were unable to compete with the low cost of FOSSIL FUELS, the Synthetic Fuels Corporation was soon disbanded. Most experts believe that synfuels will play only a minor role as an energy resource unless oil prices rise substantially in the future. [See also AIR POLLUTION; ALTERNATIVE ENERGY SOURCES; NONRENEWABLE RESOURCES; and RENEWABLE RESOURCES.]